TABLE OF CONTENTS

ACKNOWLEDGEMENT

This book would not have come to life without the support, patience, and encouragement of many people. Their guidance helped shape every idea, every explanation, and every chapter into something clearer and more meaningful. Although the subject of quantum bionics is vast and often complex, the people around me made the process feel steady, inspired, and possible.

Heartfelt appreciation goes to the scientists, researchers, and innovators whose work in quantum physics, medicine, and biotechnology continues to push the boundaries of what humanity can achieve. Their discoveries opened the door to this topic and provided the foundation on which this book stands. Their dedication to understanding the smallest parts of nature inspires hope for a healthier and more capable future.

Gratitude is also owed to those who offered feedback, shared insights, or asked thoughtful questions along the way. Every conversation helped refine the ideas and make the explanations easier to understand. Their curiosity helped shape the tone of this book, reminding me that even the most advanced concepts should remain accessible and engaging for all readers.

A special thank-you goes to the readers who pick up this book with an open mind and a sense of wonder. Their interest gives purpose to the work and helps bring these ideas out of the scientific world and into everyday thought. This book is written for them—for anyone who believes that the future of the human body holds more possibility than limitation.

Finally, thank you to the quiet moments, the late hours, and the steady motivation that carried this project forward. Every chapter is a reflection of commitment, learning, and the desire to explore a world where science can genuinely improve human life.

To everyone who supported this journey, directly or indirectly, your presence made this work stronger. This book is a tribute to your influence and to the extraordinary future that human curiosity continues to build.

INTRODUCTION

The human body has always been a source of wonder. It grows, heals, adapts, and survives in ways that have inspired scientists for centuries. Bones repair themselves, skin renews, and the brain learns new skills throughout a lifetime. Even with these remarkable traits, the body also has limits. Injuries can leave lasting damage, diseases can weaken organs, and aging gradually slows every system. For a long time, these limitations were accepted as natural boundaries. Medicine tried to treat them, and technology tried to support them, but neither could truly overcome them. Today, however, a new field is rising that may reshape the future of human health. This emerging field, known as quantum bionics, combines physics, biology, and advanced engineering to rebuild the human body in ways once thought impossible.

Quantum bionics begins with a simple idea: every part of the human body, from organs to cells, is built upon tiny particles that follow the laws of quantum physics. These particles move, react, and connect in ways that classical physics cannot fully explain. They can appear in more than one state at the same time, pass through barriers, and exchange energy across microscopic distances. This mysterious behavior is no longer only a topic for theoretical science. It is now becoming the foundation for a new era of medical innovation. When the building blocks of life are studied through a quantum lens, the body becomes something more than a biological system. It becomes an energy system shaped by patterns, fields, and interactions at the smallest scales.

The purpose of quantum bionics is not only to replace body parts with machines, but to rebuild biological functions using the same principles that form the universe. Traditional prosthetics often rely on motors, wires, and mechanical parts. Quantum-based systems use signals, energy patterns, and microscopic structures that behave more like living tissue. Instead of forcing the body to work with artificial parts, quantum bionics creates tools that feel natural to the body. These tools communicate with cells, respond to electrical impulses from nerves, and adapt to environmental changes. The focus shifts from simply repairing damage to restoring harmony within the body's natural systems.

At the center of quantum bionics lies an important concept: the body is a network of information. Each cell communicates through tiny electrical charges, vibrations, and energy exchanges. These signals shape growth, healing, and behavior. When a bone breaks, signals trigger repair. When the immune system detects a threat, signals spread to prepare a defense. When the brain forms a memory, electrical and chemical patterns form pathways that allow the information to be stored. Quantum bionics uses this language of information to design solutions that rebuild the body from within. Instead of replacing damaged structures with synthetic materials, this field aims to activate the body's own ability to heal, enhance, or adapt.

Technology inspired by quantum mechanics could lead to limbs that move as smoothly as natural arms and legs, artificial organs that never wear out, and healing systems that work faster than anything seen before. In many laboratories around the world, scientists are experimenting with quantum sensors that can detect the smallest environmental changes. These sensors might one day help the body detect diseases long before symptoms appear. Quantum materials are being developed to mimic the flexibility, strength, and responsiveness of living tissue. These materials can interact with light, sound, and electromagnetic fields in ways that open new possibilities for medicine.

Another area gaining attention is quantum communication within the nervous system. The brain processes vast amounts of information every second, and some researchers believe quantum effects may be involved in this process. If true, it may become possible to design devices that connect directly to the brain without invasive procedures. Such devices could help restore lost senses, strengthen memory, or even support people with neurological conditions. The idea of a mind working together with quantum systems sounds like science fiction, but it is increasingly being explored through real scientific studies.

The promise of quantum bionics extends far beyond healing injuries or treating diseases. It opens the path toward enhancing human abilities. Muscles strengthened with quantum materials may enable more accurate movement and less fatigue. Bones reinforced with advanced structures could become resistant to fractures. Skin designed with quantum-responsive layers may be able to sense temperature changes, pressure differences, or environmental hazards. The immune system might one day use quantum tools to identify new viruses instantly and respond quickly before illness develops.

Along with these exciting possibilities, there are also important questions. The idea of rebuilding the human body with physics challenges traditional ideas about health, identity, and the nature of being human. Some may wonder whether enhanced abilities might create a divide between individuals who have access to such technologies and those who do not. Others may worry about safety, long-term side effects, or the ethical concerns of deeply modifying biological systems. These questions must be studied with care, understanding, and responsibility. Every innovation in quantum bionics must balance progress with respect for human life.

Despite these challenges, one clear truth remains: quantum bionics represents one of the most promising directions for the future of medicine and human development. It stands at the meeting point of biology, engineering, and physics, combining them into something entirely new. The study of quantum particles is not only about understanding the universe at a deeper level. It is also about using that knowledge to create a healthier, stronger, and more adaptable human body. This field encourages imagination, pushes the boundaries of science, and offers hope to millions of people facing physical challenges.

This book explores the world of quantum bionics through simple explanations, engaging ideas, and clear descriptions. Each chapter examines how quantum principles may change the body's systems, from the senses and muscles to the brain and immune defenses. The goal is not to overwhelm with technical detail, but to open a window into a future where the human body and quantum physics work together. The journey begins with the roots of quantum bionics, moves through its present developments, and looks ahead to the possibilities waiting in the years to come.

As research continues to expand, the field grows closer to transforming medicine from treatment-focused care to enhancement-focused innovation. Quantum bionics shows that the human body is not a fixed structure with permanent limitations. It is a flexible system that can be supported, strengthened, and reshaped with knowledge drawn from the deepest levels of reality. The path toward rebuilding the human body with physics is complex, yet filled with opportunity. This introduction marks the starting point of a larger story—a story of human progress shaped by the smallest particles of the universe.

CHAPTER 1

THE BIRTH OF QUANTUM BIONICS

Quantum bionics is a new field that uses the smallest building blocks of nature to help scientists rebuild and improve the human body. While medicines, surgeries, and devices have saved millions, they cannot fully mimic how living cells work. Quantum bionics seeks better solutions by merging quantum physics—the science of atoms and particles—with advanced biomedical engineering to design systems that function more like living tissue.

Quantum physics is often called 'strange' because it behaves differently from everyday life. In the quantum world, particles can exist in more than one state at once, pass through barriers that should block them, or connect with others in ways that let information travel instantly across short distances. These behaviors sound unreal, but have been proven by experiments. Similar processes may also occur in the body, such as during protein folding, cell reactions, and neural communication. Quantum bionics uses these findings as a basis for creating new medical technologies that work in harmony with the body's natural functions.

Quantum bionics aims to repair and enhance the body by focusing on energy, information, and microscopic interactions. Unlike classical methods that support a broken bone with metal, quantum-inspired materials respond to the body's own signals to promote natural healing. This approach centers on creating solutions that work like living tissues, not just as mechanical replacements.

One key principle of quantum bionics is seeing the body as an information network. Here, cells communicate using more than simple chemistry—they use electrical charges, vibrational energy, and tiny light signals to coordinate body processes. Quantum bionics takes inspiration from this communication, aiming to develop tools that repair or improve the body with high precision.

Biomedical engineering has already made impressive progress with prosthetics, artificial organs, and implants. However, most current devices rely on mechanical or electronic systems that do not fully match the body's sensitivity. Quantum bionics pushes this progress further by introducing materials and devices that interact with the body on the same microscopic scale as cells. Quantum sensors can detect minuscule variations in temperature, pressure, and chemical concentrations, making future medical devices more responsive. Quantum materials can store or transmit energy in ways that mimic natural biological processes, giving artificial organs or prosthetic limbs a more lifelike quality.

For example, a prosthetic arm designed with quantum ideas may sense nerve signals more accurately, allowing smoother movement. The material could react instantly to the user's intent, creating a more natural feel than a mechanical one. In another case, quantum nanotechnology might one day repair damaged DNA by guiding individual atoms, lowering genetic disease risks. These concepts are not just fantasies; scientists are already exploring them in labs to see how quantum principles can improve medical technology.

Combining quantum physics and biomedical engineering paves the way for new innovations in medicine. Quantum insights explain nature's smallest mechanisms, while engineering creates practical solutions. Their partnership could lead to organs that repair themselves, skin that heals without scars, and smarter immune systems—all results of viewing the body as an information-driven system.

Quantum bionics is about creating devices and solutions that work in harmony with the human body. Rather than making biological systems adapt to artificial parts, it adapts technology to fit natural rhythms. This reduces stress, prevents rejection, and helps the whole body work together for healthier outcomes.

Another important aspect of this field is its potential to treat conditions that traditional medicine cannot. Neurological disorders, degenerative diseases, and injuries that affect the spine or brain often resist standard treatments. Quantum-level communication may help scientists design interfaces that restore lost senses or reconnect damaged neural pathways. Because the brain is believed to use extremely fine electrical and molecular signals, quantum tools could one day match these signals with precision, helping patients regain control that was once thought impossible.

Quantum bionics encourages seeing the body as an interconnected system shaped by energy and information, rather than just a collection of parts. Each function involves complex microscopic interactions. By understanding these, scientists can create solutions that respect this complexity and add support where needed.

Quantum bionics holds promise as researchers make new discoveries. Quantum sensors, dots, advanced nanomaterials, and energy systems are being tested in labs. These advances suggest a future in which diseases are detected earlier, injuries heal faster, and human biology becomes more flexible. More research is needed before such innovations are common in hospitals, but current progress shows how powerful this science partnership can be.

Quantum bionics, though in its early stages, clearly aims for a future in which human bodies can be precisely repaired, enhanced with intelligent devices, and strengthened by technology inspired by fundamental natural principles. This field brings hope to those facing physical challenges and envisions medicine with expanded possibilities. As collaboration between physics and biomedical engineering grows, the body shifts from a fragile entity defined by limitations into a dynamic system renewed, adapted, and advanced by quantum science.

History of human augmentation: from prosthetics to nanotechnology to quantum-level manipulation.

The story of human augmentation stretches back thousands of years, long before modern science or advanced machines existed. Early humans created simple tools to survive, but over time, these tools evolved into devices that supported or replaced body parts. The earliest prosthetics were made of wood, leather, and metal. These early devices were not sophisticated, yet they revealed something important about human nature. People have always searched for ways to restore lost abilities, strengthen weak functions, and overcome physical limits. A wooden leg or a basic hand might seem crude today, but in earlier times, these inventions allowed individuals to walk, hold objects, and live with dignity after injuries that would otherwise have permanently changed their lives.

As centuries passed, the craftsmanship behind prosthetics improved. Blacksmiths and early medical practitioners experimented with joints, hinges, and straps to create devices that moved with the body more naturally. Even though the materials were simple, each improvement reflected a growing understanding of anatomy and mechanics. This period showed the first major shift in human augmentation: the move from basic replacements to devices designed with the body's motion and comfort in mind. The progress was slow, but each step added something valuable to the future of biomedical technology.

The industrial age brought new materials and manufacturing methods that transformed prosthetic design. Metal alloys, springs, and molded components made artificial limbs stronger, lighter, and more responsive. Factories enabled parts to be produced with greater accuracy, providing users with more reliable support. Soldiers who lost limbs in war received improved prosthetics, which spurred further innovation. Engineers learned to create limbs that could bend, rotate, and support weight more effectively. In this era, augmentation shifted again. It was no longer only about replacing lost function; it became about restoring movement with greater precision and comfort.

As technology advanced, the idea of connecting prosthetics to the body's internal signals became a major goal. This led to the development of myoelectric prosthetics in the twentieth century. These devices used electrical signals from muscles to control artificial hands and arms. When a user attempted to move a missing limb, the muscle activity around the area created signals that the prosthetic could detect. This allowed the artificial limb to respond almost instantly. It marked a turning point because the device was no longer separate from the body—it was interacting with it. Augmentation moved from mechanical solutions to systems that relied on biological communication.

The rise of computers and electronics opened even more possibilities. Microprocessors enabled prosthetics to adapt to different walking speeds or terrain conditions. Artificial joints became smarter, learning movement patterns and adjusting support as needed. This level of intelligence made augmentation feel more natural, as if the device and the user were working together rather than separately. It also paved the way for technologies that blended engineering with deeper biological understanding.

Nanotechnology later introduced an entirely new stage in human augmentation. Instead of large mechanical structures, scientists began exploring microscopic tools that could work inside the body. Nanomaterials showed incredible strength and flexibility. They could be designed to interact with cells, proteins, and tissues with great precision. Nanobots, although still in early development, began to represent a future where microscopic machines repair damaged tissue, clean arteries, or deliver medicine directly to problem areas. This change represented a shift from external augmentation to internal enhancement. The focus moved from replacing limbs to repairing cells, improving healing, and supporting the body at a molecular level.

Nanotechnology also inspired ideas about strengthening bones, muscles, and organs using engineered materials that fuse with living tissue. Structures like carbon nanotubes and advanced polymers offered new ways to reinforce the body without heavy or uncomfortable components. Skin regeneration, nerve repair, and targeted drug delivery all benefited from nanoscale research. This period showed that augmentation did not have to be visible or mechanical. It could work quietly inside the body, improving health from within.

With these technological layers already in place, researchers began looking even deeper—into the quantum world. Quantum-level manipulation represents the newest and most ambitious chapter in human augmentation. Instead of working with molecules or cells, this approach examines the tiny particles that form all matter. Quantum physics reveals phenomena that do not occur in everyday life, such as particles occupying multiple states simultaneously or passing through barriers. While these ideas were once considered purely theoretical, they are now becoming tools for future medical innovation.

Quantum bionics aims to use these principles to design materials, sensors, and systems that behave more like living tissue than any mechanical device could. Quantum sensors may detect early signs of disease long before symptoms appear. Quantum materials may change their properties in response to the body's energy patterns, helping tissues heal faster or resist damage. Neural interfaces that operate at quantum scales may communicate with the brain with a level of accuracy that traditional electronics cannot achieve.

This newest stage of augmentation is not just about improving the body but understanding it at the deepest level. Quantum-level manipulation suggests a future where healing is faster, artificial organs are nearly indistinguishable from natural ones, and the boundaries of physical ability can be extended safely. The shift from wooden limbs to quantum systems shows how far human augmentation has come and how each generation of technology has carried forward the same basic idea—the desire to support, restore, and enhance life.

From simple prosthetics to nanotechnology and now quantum strategies, the history of human augmentation reflects constant creativity and persistence. Each era built on the challenges and successes of the previous one. The journey continues as scientists explore new ways to work with the body rather than against it, developing solutions that respect both biology and physics. What began as a struggle to replace missing limbs has grown into a field that may one day strengthen every part of the human body, inside and out, through methods shaped by the smallest components of nature.

Why classical biology is limited and how quantum principles offer breakthroughs.

Classical biology explains life through visible structures and chemical reactions. It studies organs, tissues, cells, and the molecules inside them. This approach has guided medical science for centuries and helped people understand how the body works, how diseases spread, and how injuries heal. Yet classical biology focuses mainly on what can be seen or measured at large scales, which creates natural limits. Many processes inside the body occur faster, more quickly, and more delicately than classical biology can fully describe. When reactions occur inside cells, they often follow rules that look unpredictable or unusually complex. These gaps show that something deeper influences life, something operating beyond the reach of classical explanations.

At the molecular level, classical biology often assumes that reactions follow fixed paths and predictable cause-and-effect patterns. However, living systems behave differently from most classical systems. Enzymes speed up reactions at rates far beyond what traditional chemistry can explain. DNA molecules repair themselves with striking accuracy, even though they work in an environment full of noise and constant interaction. Nerve cells pass information with remarkable speed and precision, sometimes faster than chemical reactions alone would allow. These mysteries suggest that something more subtle supports these processes—something that cannot be fully captured by classical laws.

This is where quantum principles begin to offer new insight. Quantum physics describes how particles behave at the smallest scales, such as electrons, photons, and atoms. These particles do not move in predictable paths like larger objects. Instead, they behave in ways that allow superposition, tunneling, entanglement, and rapid energy exchange. While these behaviors may seem unusual, they share many features with biological systems. Hydrogen atoms in DNA can shift positions through quantum tunneling, influencing how genes behave or mutate. Proteins fold into precise shapes through interactions that appear too quick and too organized to be random. Even photosynthesis, one of the most important biological processes on Earth, uses quantum coherence to transport energy more efficiently than any classical model can describe.

Classical biology often treats cells like small chemical machines, but this view does not fully capture the elegance of nature. Cells operate as organized networks in which information moves almost instantly, even over long molecular distances. Classical models struggle to explain how these signals remain accurate in such a crowded, constantly changing environment. Quantum principles reveal that particles can share information across tiny distances through entanglement-like effects, creating a level of coordination that classical theories cannot account for. This idea suggests that cells may use quantum behavior to maintain order and efficiency.

Another limitation of classical biology is its focus on physical structure instead of energetic patterns. Biological systems are full of subtle electrical fields, weak magnetic signals, and natural patterns of light known as bio-photons. These tiny energy flows influence how cells communicate, grow, and repair themselves. Classical models often ignore these signals because they are too faint or too fast to measure easily. Quantum physics, however, provides the tools needed to understand these subtle interactions. It shows how small changes in energy levels can trigger major shifts in biological behavior.

Quantum principles also allow scientists to imagine new kinds of medical solutions. Classical biology repairs the body by adding or removing physical material—such as drugs, implants, or external devices. Quantum-inspired approaches aim to repair the body by influencing energy states, restoring balance at the smallest scales, and guiding natural healing. Instead of forcing the body to adapt to a mechanical device, quantum systems can connect with the body's rhythms and respond instantly to changing conditions. This deeper level of communication could help repair tissues more accurately, regulate cell activity more smoothly, and prevent diseases before they appear.

One of the clearest examples is in neuroscience. Classical biology describes the brain as a network of electrical and chemical signals, but this model does not fully explain consciousness, memory formation, or the speed of thought.

Neurons process information far faster than chemical reactions allow, suggesting that extremely small and coordinated processes support brain activity. Quantum theories suggest that tiny particles inside brain cells use quantum effects to help process signals, enabling faster, more efficient communication. If this is true, future medical devices could operate at the same scale, restoring function in cases of injury or illness with far greater precision.

Quantum principles also push past classical limits in understanding diseases. Some illnesses begin with disruptions in molecular energy states rather than visible structural damage. These early changes are too subtle for classical tools to detect. Quantum sensors, however, can read energy patterns at extremely small scales. They may one day detect diseases years before symptoms appear, giving doctors the chance to treat problems long before they cause harm.

Classical biology built the foundation of modern medicine, but quantum principles offer the next step forward. They show that life is not just a collection of chemical reactions but a dynamic system shaped by energy, vibration, and subtle interactions. This more profound perspective clarifies why the body can execute intricate activities more quickly and accurately than traditional models. By using the tools of quantum science, researchers can design technologies that work with the body instead of limiting themselves to traditional methods. Quantum principles open a world of possibilities beyond the reach of classical biology, revealing new ways to repair, protect, and advance the human body.

Overview of how quantum particles, coherence, and energy states could repair or rebuild the human body.

Quantum particles behave in ways that seem unusual compared to the world that can be seen and touched. They move with incredible speed, shift between energy states, and interact through patterns that do not follow ordinary rules. These features, once the domain of physics laboratories, are now inspiring ideas for rebuilding the human body with far greater precision than traditional medicine allows. When scientists study how quantum particles organize themselves, transfer energy, or create harmony across tiny distances, they begin to see how similar behaviors could support healing inside living tissues.

One of the most promising concepts is quantum coherence. In simple terms, coherence means that particles move in a coordinated way, almost like dancers following the same rhythm. In the human body, many natural processes exhibit this type of coordination. For example, photosynthesis in plants uses coherence to transport energy extremely efficiently, allowing sunlight to be converted into useful power without wasted motion. If a similar pattern could be induced in human cells, energy could flow more smoothly through damaged tissues, speeding repair and reducing stress on surrounding structures.

Coherence may also help guide tissue repair by maintaining molecular reactions in synchrony. Often, injuries disrupt the natural order within cells, weakening or scattering signals. A coherent quantum state could restore order by helping molecules communicate in a unified pattern. This type of coordination could allow enzymes to work more effectively, guide proteins to fold into the correct shapes, and help DNA repair mechanisms fix errors with greater accuracy. Instead of relying only on chemical treatments, healing could be supported by restoring the energetic harmony that cells naturally depend on.

Another important idea is the role of energy states. Every atom and molecule in the body exists in a certain state of energy, and these states can shift under the right conditions. When energy levels are balanced, cells operate smoothly. When they fall out of balance, illness or aging may accelerate. Quantum science shows that energy levels can shift quickly and precisely, allowing certain processes to activate at the right moment. If medical technology could influence these tiny energy states, it might help stabilize damaged cells before they break apart or encourage their regeneration more effectively.

Quantum particles also move in ways that ordinary particles cannot. They can get around obstacles that would otherwise stop them by using a technique called tunneling. In the body, some reactions rely on this ability, allowing particles to move from one area to another without using much energy. This same property may one day help repair molecules inside cells by allowing specific particles to reach places that are typically difficult to access. Tunneling could also help deliver targeted treatments deep inside tissues without harming surrounding areas, improving the precision of medicine.

These quantum features are closely linked to another important quality: sensitivity. Quantum particles respond to even the smallest changes in their environment. A slight shift in temperature, pressure, or electromagnetic patterns can influence their behavior. This sensitivity could be used to identify problems inside the body long before they grow into serious conditions. Quantum sensors are being developed to detect tiny energy shifts in cells during the early stages of disease. With this technology, it may become possible to identify issues such as inflammation, infection, or cancer many years before current tools can detect them. Early detection gives the body a greater chance to heal naturally or with minimal intervention.

When these ideas come together—coherence, energy states, tunneling, and sensitivity—they create a picture of how quantum principles might rebuild the human body from the inside out. Instead of relying on large-scale surgeries or chemicals that affect the entire system, treatments could work at the smallest scale, correcting problems where they begin. Damaged cells could be encouraged to reorganize themselves by restoring their natural energy patterns. Weak tissues could be strengthened not by adding external material but by guiding internal structures back into alignment. Healing would become a process of restoring balance rather than replacing parts.

Quantum-inspired materials could also play an important role. These materials respond to energy in ways that mimic the behavior of living tissue. They can absorb and release energy, shift shapes, and interact with surrounding molecules. When used to support damaged areas, they may help guide cells into forming new structures. A damaged bone, for instance, might heal faster if supported by a quantum-active scaffold that sends signals encouraging the growth of strong, well-organized tissue. Muscles and nerves could benefit from similar guidance, helping them reconnect more smoothly after injuries.

Another possibility involves quantum communication. At small scales, particles can influence one another instantly through patterns that behave like shared information. If medical researchers learn to harness this type of interaction, they may develop healing systems that communicate directly with cells, guiding their activity without the need for invasive procedures. This type of communication could help regulate the immune system, reduce inflammation, or accelerate recovery from damage.

Through these ideas, quantum science presents a future where the body can repair itself with greater support from advanced technology. Healing becomes more efficient when the smallest components of nature are involved. The body already uses many of these principles, even if they are not fully understood. By studying how quantum behavior appears in natural biological processes, scientists can design methods that strengthen what the body already does well.

All of these concepts point toward a new form of medicine where quantum particles and energy patterns help the body rebuild with remarkable accuracy. Instead of being limited by classical approaches, healing may someday begin with the deepest layers of nature, allowing cells, tissues, and organs to restore themselves through guided energy and precise molecular coordination. The potential is vast, and as research continues, the connection between quantum behavior and biological repair becomes clearer, opening a future where the body and physics work together in ways once thought impossible.

CHAPTER 2

THE QUANTUM BODY – BIOLOGY BEYOND THE CLASSICAL WORLD

The human body is usually described in terms of cells, organs, and chemical reactions, yet beneath these familiar structures lies a much smaller and more dynamic world. At this deeper level, the body behaves less like a machine and more like a complex quantum system. Quantum physics, which studies the behavior of the smallest particles in the universe, reveals patterns that look surprisingly similar to those found in living tissues. When these patterns are examined closely, a new understanding of human biology begins to form—one that highlights the roles of energy fields, quantum tunneling, and entanglement in the processes that sustain life.

Every cell in the body is surrounded by an invisible field of energy. This field is created by tiny electrical charges moving across cell membranes, by the vibrations of molecules, and by the natural signals generated inside tissues. These energy fields are not random; they help guide cell communication and influence how the body responds to stress, injury, or illness. Classical biology has studied electrical activity for years, but quantum science shows that these fields may operate with far greater sensitivity than previously believed. They can shift in response to extremely small influences, meaning even subtle changes in the body's environment can affect cellular behavior. In this sense, the body acts like a finely tuned quantum network in which energy, rather than just chemistry, helps organize life.

One of the most fascinating features of quantum physics is tunneling, a phenomenon in which particles can pass through barriers that should block them. On a large scale, this makes no sense, but at the smallest level, it happens all the time. Inside the human body, tunneling plays a key role in many biochemical reactions. Enzymes, which control nearly every process from digestion to DNA repair, work far faster than classical chemistry predicts. This speed is partly explained by tunneling, which allows particles to jump to new positions without following the slower, step-by-step paths expected in classical systems. Through tunneling, reactions that would take hours or days occur in fractions of a second, allowing the body to adapt quickly and efficiently.

DNA itself benefits from quantum tunneling. The hydrogen atoms that help form DNA's structure can shift positions through tunneling, influencing how genetic information is stored, repaired, or expressed. This tiny shift can change the shape of DNA or correct errors that might otherwise lead to mutations. Classical biology describes DNA as a long chain of molecules, but quantum mechanics shows that this chain is dynamic, with even the smallest particles contributing to its stability and function. This hidden activity supports the idea that life relies on quantum events to maintain its integrity.

Another key quantum concept found in nature is entanglement. In physics, entangled particles share a connection that allows them to influence each other instantly, even when separated. While this behavior seems impossible in ordinary life, signs of similar coordination appear in biological systems. For example, energy transfer during photosynthesis exhibits patterns similar to those of entanglement, enabling energy to move with almost perfect efficiency. In the human body, molecules involved in sensing, signaling, and tissue repair may also engage in coordinated interactions that resemble quantum entanglement. This helps explain how cells can communicate so quickly across distances that are too large for ordinary molecular contact.

The nervous system provides another example. Neurons send signals using both electrical impulses and chemical messengers, but the timing is often too precise to be explained by classical processes alone. Some researchers believe that entanglement-like effects may help synchronize neural network activity. If true, this could shed light on the speed of thought, the formation of memories, and the body's ability to process complex information almost instantly. It suggests that communication in the brain is not only mechanical but also influenced by subtle quantum interactions that maintain signal organization.

Energy fields, tunneling, and entanglement work together to allow the body to function with incredible precision. These processes enable molecules to find their correct partners, cells to coordinate their actions, and tissues to heal accurately. When seen through a quantum lens, the body becomes more than a biological structure; it becomes a system guided by patterns of energy and information that stretch far beyond traditional explanations.

Understanding the body as a quantum system also changes how healing is viewed. If energy fields influence cellular behavior, then restoring balance in these fields may support recovery. If tunneling helps reactions proceed quickly, it may be possible to guide or enhance them when the body struggles to heal. If entanglement-like coordination organizes molecular communication, then disruptions in this coordination might explain certain diseases or disorders. By paying attention to quantum behavior, scientists gain new ways to think about treatment—not only to correct physical damage but also to support the underlying energetic harmony that keeps the body stable.

This view does not replace classical biology; instead, it adds depth to it. The body still relies on chemistry, structure, and genetics, but these systems operate more effectively when supported by quantum-level interactions. Energy fields help cells communicate. Tunneling speeds up essential reactions. Coordinated interactions allow tissues to stay organized even in constantly changing conditions. Together, these features create a picture of life that is far more dynamic and interconnected than traditional models suggest.

Seeing the body as a quantum system opens new possibilities for understanding health and creating future technologies. It provides a foundation for quantum bionics, where devices and materials are designed to work with the body's natural energy and communication patterns. This approach may lead to more precise treatments, faster healing, and stronger support for the body during illness or aging.

The human body is not simply a collection of organs and cells. It is a living quantum network, shaped by energy, motion, and invisible connections that operate at the smallest scales of nature. By recognizing and studying these quantum features, a clearer picture of life emerges—one that brings science closer to understanding how the body truly works and how it can be supported in ways once considered impossible.

How quantum effects already operate in DNA, enzymes, and cellular communication.

Quantum effects are often described as strange or unfamiliar, yet they quietly influence some of the most important functions inside the human body. Instead of being limited to physics experiments or advanced theories, quantum behavior appears in DNA, in enzyme activity, and in the communication systems that link cells together. These effects help life operate at speeds, with accuracies, and with efficiencies that classical biology alone cannot explain. When these hidden processes are uncovered, the body begins to look less like a collection of isolated chemical reactions and more like a finely tuned system guided by subtle patterns of energy and motion.

DNA provides one of the clearest examples of quantum activity in nature. At first glance, DNA appears to be a simple ladder-shaped molecule that carries genetic information. Classical biology describes how its structure allows cells to copy genes and repair damage. However, many of these actions rely on movements too fast and too precise to be explained by ordinary chemistry. One important quantum-mechanical phenomenon in DNA is tunneling, which allows tiny particles to shift positions without the energy normally required to cross a barrier. Hydrogen atoms in DNA often tunnel between different positions along the molecule. This helps maintain stability by supporting accurate base pairing and also helps correct errors. The body depends on this constant motion to maintain the reliability of genetic information, even as cells divide countless times throughout a lifetime.

DNA also needs to fold and twist into tight structures to fit inside each cell. These shapes change in response to small energy signals, not just chemical forces. Quantum principles help explain how DNA can respond so quickly to these signals. Even slight changes in energy levels can alter the molecule's structure, allowing it to switch between active and inactive states with impressive speed. This sensitivity helps regulate which genes turn on or off, guiding growth, healing, and daily biological functions. Without quantum effects, DNA would behave more rigidly and respond much more slowly, limiting the body's ability to adapt smoothly to changing conditions.

Enzymes, the tiny biological machines that speed up chemical reactions, also rely on quantum behavior. Enzymes work far faster than classical chemistry predicts. Many reactions that should take hours or days happen almost instantly inside the body. This efficiency is partly explained by quantum tunneling. Instead of waiting for particles to gain enough energy to cross a barrier, enzymes guide certain particles into positions where tunneling becomes likely. This allows reactions to proceed at extraordinary speed with minimal energy use. Enzymes use this property to support nearly every process in the body, from breaking down food to repairing cells and producing energy. Their remarkable speed shows that biology has evolved to take advantage of quantum phenomena whenever they improve performance.

Some enzymes also exhibit quantum coherence, a phenomenon in which particles move in a coordinated, orderly manner. Coherence allows systems to respond quickly and precisely, even in environments full of noise and motion, such as the inside of a cell. This coordinated movement explains how energy travels efficiently along enzyme structures, guiding reactions in a controlled way. When coherence is lost, reactions slow down or become less accurate. The presence of coherence inside enzymes suggests that the body uses quantum rules not only for speed but also for stability and control.

Cellular communication is another area where quantum effects appear to play a major role. Cells constantly exchange signals to maintain balance and respond to changes. These signals involve electrical charges, light particles, and small shifts in energy patterns. Some of these signals are extremely faint, yet cells can detect them with high accuracy. This level of sensitivity suggests that quantum processes may help amplify or organize these signals.

Bio-photons—tiny particles of light produced by cells—are a good example. These light signals are extremely weak, yet they appear to help coordinate cellular activity. Their behavior resembles quantum interactions because they travel in patterns that remain organized rather than scattering randomly. This organization allows cells to communicate quickly and maintain harmony across tissues.

Neurons may also use quantum behavior. The brain processes enormous amounts of information with remarkable speed, and some researchers suspect that quantum effects contribute to the timing and accuracy of neural activity. Certain structures inside neurons, such as microtubules, show properties that could support quantum interactions. These structures may help synchronize signals or support memory formation by enabling extremely rapid energy changes. Although research is still ongoing, the evidence suggests that the brain may rely on quantum behavior to achieve the level of precision needed for complex thought and perception.

Another example is the way cells sense tiny changes in their environment. Many biochemical processes operate at the edge of what classical theories can describe. For instance, some proteins change shape in response to extremely small energy shifts—far smaller than what ordinary physics predicts they should notice. Quantum sensitivity helps explain how cells detect signals such as magnetic fields, temperature changes, or chemical concentrations that are too subtle for classical sensors.

These examples show that quantum effects are not foreign to biology. They operate quietly within DNA, support enzyme activity, and help cells communicate with extraordinary precision. Instead of seeing life as a set of rigid mechanical systems, the body emerges as an adaptable network shaped by motion, energy, and delicate interactions. The more researchers learn about these hidden behaviors, the more they understand how the body achieves such efficiency and resilience.

Recognizing quantum influences in biology also opens new possibilities for medicine. If tunneling speeds up reactions, future treatments could use similar principles to support healing. If coherence helps organize molecular activity, therapies could aim to restore coherence when it is disrupted. If cells rely on energy sensitivity, medical devices could be designed to support these signals rather than overwhelm them. Quantum behavior endows the body with abilities that classical biology cannot fully explain, and these abilities may hold the key to stronger, more precise approaches to health.

The role of bio-photons and electromagnetic signatures in healing.

Inside every living cell, tiny particles of light are constantly being produced and exchanged. These particles, known as bio-photons, are incredibly faint—far too weak for the naked eye to see. Yet their presence reveals something remarkable about how the body operates. Instead of depending only on chemical reactions, the body also communicates and organizes itself through subtle light signals and electromagnetic patterns. These delicate forms of energy help guide healing, growth, and balance in ways classical biology has only recently begun to understand.

Bio-photons are generated by various molecular processes, especially those involving oxygen and energy production. Every time a cell breathes, creates fuel, or repairs itself, small bursts of light are released. These light emissions are not random sparks. They follow patterns linked to the body's internal rhythms. Healthy tissues produce stable, organized biophoton signals, while stressed or damaged tissues exhibit irregular or scattered patterns. Researchers who study these signals have found that their coherence—how smoothly they flow—often reflects the overall health of the tissue. When coherence is strong, communication inside the body becomes clearer, allowing cells to coordinate healing more effectively.

The idea that light inside the body supports healing may sound surprising, but it becomes easier to understand when seen as a communication system. Bio-photons travel between cells and carry information about activity, stress levels, or needed repairs. This helps cells adjust their behavior based on their neighbors' behavior. When tissue is damaged, surrounding cells change their bio-photon patterns, signaling that help is needed. Nearby cells respond by sending energy, activating growth pathways, or increasing repair activity. This process happens quietly and continuously, forming a subtle network of light-based communication.

Another important factor is the electromagnetic signature that surrounds every cell. Electrical charges move across cell membranes, creating tiny electromagnetic fields. These fields shift constantly as cells divide, process energy, or respond to signals. When many cells work together, their individual fields can merge into larger patterns that guide tissue behavior. For example, bone cells use electromagnetic cues to determine how to rebuild a fracture. The heart creates strong electrical patterns that coordinate each beat. Even the brain relies on delicate electromagnetic rhythms to form thoughts, memories, and sensory experiences.

During healing, these electromagnetic signatures play a major role. When tissue is injured, the normal electromagnetic pattern becomes disrupted. This disruption sends a clear signal to the surrounding area, similar to an alarm. The body responds by directing resources toward the injury: white blood cells arrive, new blood vessels form, and damaged cells are cleared. The electromagnetic change acts like a map that helps direct all these activities. As healing progresses, the field gradually returns to its normal state, signaling that the repair process is complete.

This connection between energy and healing becomes even clearer in studies of tissue regeneration. Experiments have shown that electromagnetic stimulation can accelerate wound healing, promote bone growth, and reduce inflammation. Low-frequency electromagnetic fields can help align cells, making them work in harmony. These fields appear to support recovery by strengthening the body's natural energy patterns. Instead of forcing healing from the outside, they help restore the internal order that guides recovery.

Bio-photons also play a role in regulating the balance between oxidative stress and cellular protection. Their emission levels often rise when cells are under strain, providing a subtle indicator that something is wrong. This allows nearby cells to respond before damage becomes severe. In this way, biophotons help maintain balance across tissues. They allow the body to sense small changes quickly, long before symptoms become noticeable. This sensitivity may explain why some healing processes begin almost instantly after an injury.

Another fascinating aspect is that certain therapies used in medicine may work, in part, by influencing the body's electromagnetic or photonic signals. Techniques like laser therapy, pulsed electromagnetic therapy, or even exposure to natural sunlight can affect how cells emit bio-photons or shape their electromagnetic patterns. These treatments do not just act on the physical structure of tissues; they also help restore the energetic environment needed for efficient healing. When the energy patterns around cells are stable and clear, cells perform their roles with much greater precision.

The presence of bio-photons also raises the possibility that cells share information faster than chemicals alone would allow. Chemical signals can be slow because they must travel across space and bind to receptors. Light signals, however, move at incredible speeds. This suggests that biophotons may serve as a rapid communication mechanism, providing cells with immediate updates on the condition of their environment. If cells can respond quickly, healing can begin sooner, and the overall process becomes more organized.

Electromagnetic signatures add another layer to this communication system. Cells generate these patterns not only during repair but also during everyday activities. When the patterns become chaotic, it can signal stress or illness.

Returning these patterns to harmony may help restore health. This idea is supported by the observation that meditation, controlled breathing, and relaxation techniques can influence the body's electromagnetic rhythms, especially in the heart and brain. These practices often lead to improved recovery, reduced pain, and better immune function.

Understanding bio-photons and electromagnetic signatures does not replace classical biology. Instead, it gives a more complete picture of how healing works. Chemical reactions provide the raw materials for repair, while quantum-level signals guide the timing, coordination, and direction of these repairs. Together, they form a system that is both physical and energetic.

The body heals not only through physical reconstruction but also through these subtle forms of communication. Bio-photons carry light-based information, while electromagnetic signatures shape the environment in which cells operate. Both are essential for maintaining order and guiding recovery. When these signals flow smoothly, healing becomes more efficient, and the body's natural resilience becomes stronger.

Why understanding the quantum body is key to creating advanced bionic solutions.

Creating advanced bionic solutions requires more than understanding bones, muscles, and organs in a mechanical way. The body is not just a structure made of tissues; it is a living network shaped by tiny particles, energy patterns, and microscopic interactions governed by quantum physics. When these hidden processes are ignored, bionic devices often feel unnatural or fail to match the speed and precision of real biological functions. Understanding the quantum body provides engineers and scientists with a deeper foundation for designing solutions that communicate with the body at the same level that cells, DNA, and enzymes naturally use. This deeper understanding is what makes the next generation of bionics possible.

Traditional bionic devices rely on electrical signals and mechanical parts, which can only imitate biological movement. They respond to muscle signals but do not fully reflect how cells coordinate actions or how energy flows through tissues. The human body works with remarkable accuracy because much of its communication happens through quantum processes. Reactions inside cells depend on extremely small energy shifts and rapid exchanges of information. If bionic systems do not recognize these quantum patterns, they remain limited. They can restore basic movement, but they cannot achieve the natural feel, fluid motion, or real-time responsiveness of living limbs and organs.

When the quantum body is understood, bionic devices can be designed to work with the body rather than beside it. For example, enzymes speed up reactions through quantum tunneling, allowing the body to perform tasks with very little energy. A bionic device that understands this principle could operate in harmony with the body's chemistry, using minimal power while delivering fast results. Instead of relying on batteries or heavy machinery, future bionics could shift energy states the same way cells do, creating devices that feel alive and responsive.

Another important part of the quantum body is coherence, in which particles move in an organized, synchronized way. Coherence helps support processes such as healing, sensing, and energy transfer. When tissues work together through coherent quantum patterns, they produce smooth and accurate responses. If bionic technologies can recognize and support this coherence, they will be able to integrate more naturally. A limb controlled through quantum-aligned signals could match the rhythm of nerve impulses far better than one using simple electrical detection. A bionic heart built with quantum-sensitive materials could adjust its beat instantly based on subtle energy shifts, giving it a level of coordination no mechanical pump can match.

Quantum communication also plays a silent but powerful role inside the body. Cells respond to extremely faint electromagnetic signals and tiny bursts of light known as bio-photons. These signals help guide repair, regulate growth, and maintain balance. A bionic system that understands these signals can respond more naturally to the body's needs. Instead of waiting for large electrical impulses, it can detect the small changes in energy that occur before a muscle contracts or before a nerve fires. This early detection allows bionic devices to act with speed and precision that mimics natural biology.

Understanding the quantum body also gives engineers new pathways to improve sensing capabilities. Human senses rely on tiny molecules responding to light, sound, and chemical information through quantum interactions. Vision, for example, begins with a single photon triggering a reaction in the eye. Smell depends on molecules vibrating at quantum frequencies. If bionic sensors are built with the same principles, they can capture information with extraordinary sensitivity. A bionic eye designed with quantum detectors could see in extremely low light or interpret wavelengths beyond human vision. A bionic ear inspired by quantum vibrations could hear frequencies with stunning accuracy. These improvements are only possible when the underlying quantum behaviors are acknowledged and used.

The biggest breakthrough comes from understanding how the nervous system may use quantum effects. Nerve impulses are not only electrical; they also depend on small shifts in energy and structural changes inside cells. Some researchers believe certain parts of neurons may support quantum-level communication, helping the brain coordinate information far faster than classical models predict. If bionic interfaces can connect at this level, communication between the brain and artificial devices becomes smoother and more natural. Instead of struggling to interpret vague electrical signals, bionic systems could understand the deeper patterns behind intention, motion, or sensation.

Quantum knowledge also helps create materials that behave more like living tissue. Traditional prosthetics use metals, plastics, and mechanical joints. Quantum-based materials can store energy, change shape, or adapt their properties based on tiny environmental cues. These materials could make bionic limbs lighter, stronger, and more comfortable. They could adjust stiffness during movement, respond to temperature changes, and even support self-repair. The body responds best to materials that behave in familiar ways, and quantum-informed engineering brings artificial systems closer to the structure and behavior of natural tissues.

Another reason understanding the quantum body is essential is the role of energy. Biological systems move energy with incredible efficiency. Quantum coherence and tunneling help reduce waste, allowing cells to operate using only small amounts of fuel. Advanced bionic devices inspired by these principles could run longer, require less power, and perform more smoothly. Instead of forcing energy through circuits, they could use quantum pathways that mimic natural energy flow.

Finally, understanding the quantum body helps build trust between technology and biology. The body accepts systems that speak its language. When artificial devices align with quantum patterns—light signals, electromagnetic rhythms, and molecular vibrations—the immune system may show lower rejection, tissues may adapt more smoothly, and long-term use becomes safer and more reliable. The future of bionics depends on this harmony.

Advanced bionic solutions cannot reach their full potential using classical ideas alone. They need the insights that come from studying the body at its deepest level. Quantum behaviors shape how life grows, heals, senses, and communicates. When these behaviors guide engineering, the result is technology that feels natural, behaves intelligently, and supports the human body with unmatched precision. Understanding the quantum body is not just useful; it is the key that unlocks a new generation of bionic possibilities.

CHAPTER 3

QUANTUM SENSORS – RETHINKING THE FIVE SENSES

Quantum-enhanced prosthetics represent one of the most exciting steps toward creating artificial limbs that feel naturally connected to the human body. Traditional prosthetics rely on surface muscle signals or simple mechanical controls, which often limit precision and create a delay between intention and movement. Quantum technology changes this by allowing artificial limbs to sense electrical fields, magnetic waves, and even subtle molecular vibrations produced by the body. These signals are incredibly small, yet they carry rich information about intention, pressure, balance, and emotional state. When a prosthetic can read and respond to them, the result is movement that feels smoother, faster, and more lifelike.

Inside the human body, every cell generates small electrical fields as ions move in and out of membranes. Muscles, in particular, create distinct patterns when they prepare to contract. In traditional prosthetics, only the largest of these signals can be detected through electrodes placed on the skin. Much of the finer information remains hidden. Quantum-enhanced sensors, however, can detect electrical fields that are far weaker yet far more detailed. These sensors operate at a sensitivity level that allows them to read signals before the muscle even fully activates. This gives the prosthetic a head start, resulting in faster, more natural movements. Instead of waiting for forceful muscle contractions, the device responds to the earliest signs of intention.

Magnetic waves produced by neurons also hold valuable information. As electrical impulses travel through the nerves, tiny magnetic fields accompany them. These fields are too small for classical sensors to detect reliably, but quantum sensors can register them with impressive accuracy. By reading these magnetic patterns, a prosthetic can interpret messages directly from the nervous system. This creates a more seamless form of control because the limb begins moving at almost the same instant the brain sends the signal. The connection feels closer to how a natural limb behaves, reducing the strain often experienced with older prosthetic systems.

Molecular vibrations add another layer to this communication. Inside muscles, bones, and connective tissues, countless molecules vibrate at different frequencies. These vibrations shift depending on movement, tension, and emotional state. Classical technology ignores these signals because they are too subtle, but quantum-based sensors can interpret them as additional information. By reading molecular vibrations, a prosthetic can learn how much force is needed for a certain action, whether the user is tense or relaxed, and how the body is adjusting to balance. This makes the device more intuitive. It can lighten its grip when sensing strain or apply more power during motion that requires strength.

Quantum-enhanced sensors do not simply detect these signals—they interpret them as part of a continuous flow of information. This creates a rich, multilayered, and fast communication pathway between the user and the device. The prosthetic becomes an extension of the body rather than a tool attached to it. Movements that once felt difficult, slow, or awkward become smoother and more controlled. Users often describe a sense of connection, as if the artificial limb were listening to the body rather than waiting for commands.

Another powerful feature of quantum-enhanced prosthetics is adaptability. Because these sensors read such detailed signals, the device can learn patterns over time. It recognizes how the user walks, how they naturally shift weight, and how their muscles behave during everyday activities. This learning makes the prosthetic more personalized. Instead of giving the same response to every user, it adapts to each user's movement style, comfort level, and unique physical signals. This creates a sense of ease and comfort that older prosthetics cannot offer.

Some advanced designs also incorporate feedback systems that send information back into the body. Quantum materials can generate tiny electromagnetic patterns or molecular vibrations that mimic natural sensory signals. When these signals reach the nerves, the user feels sensations such as pressure, texture, and even temperature. This creates the possibility that touch may return to individuals who thought they had lost it forever. Imagine gripping an object and feeling how firm it is, or sensing the warmth of a cup—these experiences are made possible by quantum-based sensory feedback.

This deeper interaction between the body and the prosthetic also reduces fatigue. Older devices often require strong, repeated muscle signals for movement, which can lead to fatigue. Quantum-enhanced prosthetics need only the smallest signals to function. This reduces strain and encourages more fluid movement throughout the day. The limb moves when the body hints at movement, not when it forces it.

Quantum technology also opens opportunities in environments where precise control is essential, such as sports, delicate crafts, and even surgical work. The ability to sense subtle changes in balance, tension, and electromagnetic patterns enables refined motion that rivals, or even surpasses, natural movement. Instead of limiting users, a quantum-enhanced prosthetic becomes a source of confidence and expanded capability.

The emotional impact of this technology is equally important. When a prosthetic responds naturally, users often feel more whole and more comfortable in their daily lives. They regain independence, confidence, and the ability to engage in activities they once avoided. The feeling of connection reduces frustration and helps restore a sense of normalcy.

Quantum-enhanced prosthetics are more than advanced machines. They represent a new relationship between technology and the human body—a relationship built on understanding signals once considered too small to notice.

By reading electrical fields, magnetic waves, and molecular vibrations, these devices bridge the gap between intention and action. They restore movement not through mechanical strength but through harmony with the body's own language. As research continues, this technology promises a future where artificial limbs feel natural, responsive, and deeply integrated, offering possibilities once thought unreachable.

Vision upgrades using quantum dots, photon amplification, and night-vision integration.

Vision upgrades built through quantum technology open a path to sight that goes far beyond the natural limits of the human eye. Instead of relying only on biological structures, these upgrades use quantum dots, photon amplification, and advanced night-vision systems to create sharper, brighter, and more detailed visual experiences. The goal is not just to restore sight but to enhance it in ways that feel natural, effortless, and deeply connected to how the brain interprets the world.

Quantum dots play a central role in these upgrades. These tiny particles, far smaller than a cell, can absorb and release light with extraordinary accuracy. Unlike traditional materials, quantum dots respond to specific wavelengths and convert them into signals the eye or a visual device can use more effectively. When placed in retinal implants or visual lenses, they help sharpen images, boost color clarity, and make low-light environments easier to navigate. Their precision allows them to adjust instantly to changes in brightness, reducing glare during the day and brightening dim scenes at night. This creates a vision that adapts more smoothly than the natural eye, offering comfort and clarity in every environment.

One of the most impressive abilities of quantum dots is their power to enhance color range. Human eyes can see a limited spectrum, but quantum dots can interact with wavelengths beyond this range. When these particles convert invisible light into visible light, they expand the range of what the user can perceive. Subtle shifts in color that would normally go unnoticed become vivid and detailed. Landscapes appear richer, shadows hold more depth, and fine differences between similar shades become clearer. This heightened perception feels natural rather than artificial because the brain quickly adapts to the new information.

Photon amplification adds another layer to this enhanced vision. Inside biological eyes, vision depends on the number of photons available. When the light is too weak, the retina struggles, and images become blurry or grainy. Photon amplification changes this limitation by strengthening even the faintest light. Quantum sensors collect small amounts of light that would otherwise be lost, amplify them, and send stronger signals to the visual pathways. This allows clear sight in environments that would otherwise be nearly impossible to navigate. A dimly lit street, a moonlit forest, or a cloudy evening becomes visible with surprising detail.

Photon amplification also improves the sharpness of motion. Fast movements or distant objects become easier to track because the visual system receives stronger, more consistent light signals. Rather than stretching the eye's natural ability, these enhancements support it with extra clarity. This creates smoother transitions, reduces eye strain, and makes high-speed environments more comfortable.

Night-vision integration brings these benefits together. Traditional night-vision systems often rely on bulky equipment or produce green-tinted images that look unnatural. Quantum-based night-vision, however, works directly with the eye or with compact implants. It uses quantum dots and photon amplification to gather near-infrared and low-energy light, turning it into a clear, full-color image. Instead of seeing only outlines, users can see real detail: facial expressions, textures, and movement in shadowed spaces.

This integration gives night vision a natural feel. There is no sudden brightness or disorienting contrast. Light appears evenly distributed, creating an experience similar to daylight vision but softer. Walking through a dark room no longer requires caution. Outdoor environments reveal shapes, paths, and objects that would normally be hidden. For anyone who regularly relies on their vision, such as first responders, pilots, or people navigating unfamiliar locations, this smoother transition between light and darkness is extremely beneficial.

Another remarkable aspect of these upgrades is their ability to blend multiple lighting sources into a single coherent image. In natural vision, bright lights and dark spaces appearing together make it difficult for the eye to adjust. Quantum technology balances these extremes by adjusting the amount of light amplified in different parts of the scene. A bright streetlamp does not overpower nearby shadowed areas. Both can be seen clearly without discomfort. This makes vision more stable and reduces the strain caused by sudden lighting changes.

Quantum-enhanced vision also improves depth perception. Because the upgrades gather more information from each photon, the brain receives a fuller picture of distance and texture. Surfaces appear more detailed, edges more defined, and movement easier to interpret. Even individuals with limited depth perception may experience noticeable improvement, allowing them to judge distances more accurately and avoid obstacles with greater confidence.

These vision upgrades also support emotional comfort. Clear and reliable sight in all lighting conditions reduces anxiety and improves independence. Tasks that once felt difficult—driving at night, reading small text, exploring dim areas—become natural again. Vision becomes a stable part of daily life rather than a limitation or concern.

The combination of quantum dots, photon amplification, and night-vision integration represents a new era of visual enhancement. These tools do not overpower natural vision; they strengthen and expand it. They help the eyes receive more information while keeping the experience comfortable and familiar. Whether used in medical implants, smart lenses, or future bionic eyes, quantum-enhanced vision offers a brighter, sharper, and more confident view of the world.

Quantum hearing implants interpreting subatomic vibrations for ultra-sensitive sound detection.

Quantum hearing implants take the idea of restored hearing far beyond traditional devices. They do not simply amplify external sound. Instead, they interpret the smallest vibrations—movements so subtle they occur at the level of atoms—and convert them into clear, meaningful audio signals. This level of sensitivity allows users to hear details that ordinary hearing, or even classical hearing aids, could never detect. With quantum technology, sound becomes richer, cleaner, and more precise, opening a new dimension of auditory experience.

Inside the ear, sound waves move through the air, striking the eardrum and causing the tiny inner ear hair cells to vibrate. These hair cells convert mechanical motion into electrical signals that the brain interprets as sound. When these cells are damaged, hearing becomes difficult or impossible. Traditional hearing aids try to fix this by amplifying sounds. By delivering electrical pulses straight to the auditory nerve, cochlear implants avoid damaged areas of the ear. While helpful, both approaches have limitations. They struggle with soft sounds, background noise, and complex audio environments.

Quantum hearing implants transform this process. Instead of waiting for large sound waves or relying on mechanical vibrations, they detect subatomic movements in the air and inside surrounding tissues. Every sound—even the quietest whisper—creates patterns at the quantum level. Molecules vibrate, electrons shift, and tiny energy changes ripple outward. Classical devices cannot detect these faint signals, but quantum sensors can. They read these delicate patterns with extraordinary accuracy, giving users access to layers of sound that were once invisible to human perception.

One of the key strengths of quantum sensors is their ability to interpret vibrations before they become full sound waves. When a person speaks, the body produces tiny molecular changes in the air near the mouth. These changes occur faster than traditional microphones can capture. Quantum implants sense these early signatures, allowing the user to hear speech with remarkable clarity and almost no delay. This also helps with understanding conversations in noisy places. Instead of amplifying all sounds equally, the implant focuses on the earliest, cleanest parts of speech, separating them from background noise long before the sounds overlap.

The sensitivity of quantum implants also affects how users perceive low- and high-frequency sounds. Classical hearing devices often struggle with very low rumbles or extremely high tones, creating a distorted listening experience. Quantum sensors, however, read vibrations across a much wider range. They can sense deep vibrations from distant objects, subtle shifts in air pressure, and delicate harmonics in musical notes. This gives the listener a fuller, more natural soundscape. Music becomes richer, voices become warmer, and environmental sounds become more detailed.

Another advantage comes from how quantum implants interact with the body. Instead of relying solely on air vibrations, they can detect energy changes within tissues near the implant. Bones, for instance, carry vibrations extremely well. Quantum sensors pick up these internal signals and combine them with external sound information. This dual input allows for a more stable understanding of pitch, direction, and intensity. Users can pinpoint the source of sounds more accurately, even in complex environments like crowded streets or echoey rooms.

This increased awareness also offers safety benefits. Footsteps, soft mechanical hums, and slight shifts in the environment become noticeable earlier. A car approaching from behind or a machine starting up in the distance creates vibrations that a quantum implant can detect long before they register as sound. This helps people navigate the world with greater confidence, especially in situations where awareness is crucial.

Quantum implants can also adapt to the emotional content behind a sound. Human voices carry subtle frequencies connected to mood, intention, and stress. These cues are often too faint for ordinary hearing, but they appear clearly in the quantum vibration patterns around speech. The implant interprets these patterns, allowing the listener to sense emotion more accurately. This adds depth to conversations, strengthening social connections and improving communication.

Another remarkable feature of these implants is their ability to learn from the user. The brain naturally adapts to new patterns, and quantum devices are designed to adapt in return. Over time, the implant gathers information about the user's preferences—what sounds they focus on, which tones they find comfortable, and how they respond to different environments. The device fine-tunes itself to match the user's natural listening habits, creating a personalized auditory experience that feels effortless.

Quantum technology also reduces listening fatigue. Traditional devices often overwhelm the brain with amplified noise, especially in loud or chaotic places. Quantum implants do not rely on volume. They rely on detail. By interpreting precise vibration patterns, they deliver sound that feels smooth and natural, reducing strain on the auditory system. This allows users to enjoy long conversations, music, or outdoor activities without feeling overwhelmed.

In medicine, quantum hearing implants may help diagnose conditions that affect balance or neurological function. Because they detect the smallest internal vibrations, they can sense changes in how the inner ear or nerves behave. This information can help doctors identify issues earlier and provide better treatment.

Quantum hearing is not only a technological upgrade—it is a new form of listening. It reveals layers of sound that have always been present but hidden. It strengthens the connection between the ear and the brain by using the body's own quantum signals. With this approach, hearing becomes more than sensing noise; it becomes an experience of clarity, precision, and connection.

Touch, taste, and smell augmentation through quantum-level sensory mapping.

Touch, taste, and smell are often described as simple senses, yet they shape how people understand the world. They influence memory, emotion, comfort, and instinct. When any of these senses weaken or disappear, daily life loses richness. Quantum-level sensory mapping introduces a new approach to restoring and enhancing these abilities. Instead of relying on broad electrical signals or mechanical sensors, this method focuses on the smallest energy and vibration patterns that define each sensation. By understanding these quantum patterns, scientists can design systems that allow artificial skin, synthetic taste receptors, and enhanced smell detectors to communicate with the brain in a way that feels natural and immediate.

Touch begins with tiny receptors in the skin that react to pressure, temperature, and texture. Classical technology attempts to replicate this using pressure or heat sensors, but these systems lack the complexity of real skin. Touch is not only about force; it is about patterns. Every surface creates a unique vibration when fingers move across it. These vibrations reach the skin in complex waves that the brain interprets as softness, smoothness, roughness, or warmth. Quantum-level mapping captures these waves by reading the smallest changes in energy. Instead of waiting for strong pressure, quantum sensors detect the subtle shifts at the molecular level when skin and an item come into contact. This creates a version of touch that feels authentic, not mechanical.

Artificial skin enhanced with quantum sensors can respond to texture with remarkable detail. It can sense the difference between silk and cotton, between warm ceramic and cool metal. These tiny variations come from subtle differences in molecular vibrations, something classical sensors miss.

Quantum systems, however, recognize these shifts and translate them into electrical signals that the nervous system can understand. For someone with a prosthetic limb, this means feeling the world again—feeling the softness of fabric, the smoothness of a cup, or the gentle pressure of holding a friend's hand. The experience becomes emotional as well as physical, because touch carries comfort and connection.

Taste relies on chemical interactions, but these interactions begin with extremely small energy exchanges. When food touches the tongue, molecules vibrate and release signals that travel through taste receptors. Quantum mapping focuses on these early signals. It identifies the energy patterns of sweet, sour, salty, bitter, and savory flavors long before they become fully developed chemical reactions. By capturing these patterns, artificial taste receptors can recreate flavor with surprising accuracy. They do not simply measure chemical concentration; they measure the energetic signature of each taste.

This technology could help people who have lost their sense of taste due to injury or illness. With quantum sensors placed inside the mouth or integrated into a smart palate device, flavors can be detected and translated into signals sent to the brain. This creates a sensation similar to natural taste. As technology improves, these systems may even enhance taste by amplifying subtle flavor notes that natural tongues often miss. Food could become more enjoyable, more layered, and more exciting. Chefs, food scientists, and individuals with sensory impairments could all benefit from this form of enhancement.

Smell, one of the most emotionally powerful senses, also begins with quantum behavior. Many researchers believe that smell depends not only on chemical shape but also on molecular vibration. Each scent molecule vibrates at a specific frequency, and receptors in the nose respond to these frequencies. This quantum vibration theory helps explain how humans can distinguish thousands of scents with remarkable precision. Quantum-level mapping builds on this idea. Instead of simply detecting the presence of scent molecules, quantum sensors read their vibrational patterns directly. This allows a sensory device to recognize scents that would normally be too faint or too complex for classical detectors.

Smell augmentation using quantum sensors can be life-changing for people who have lost their sense of smell after infections, trauma, or aging. By identifying vibrational signatures, artificial olfactory devices can send signals to the brain that match the experience of real smell. Users may once again recognize the scent of fresh rain, familiar spices, or a loved one’s perfume. This restoration is not just practical; it brings emotional comfort. Smell plays a major role in memory, bonding, and daily enjoyment. With quantum mapping, these experiences become possible again.

Smell enhancement also opens new opportunities for safety and precision. Quantum sensors can detect dangerous chemicals, gas leaks, or spoiled food long before they reach harmful levels. They can detect smoke signatures even at early stages, offering quicker warning than traditional fire detectors. Workers in hazardous environments, emergency responders, and even households could benefit from this heightened awareness.

Another interesting possibility is the integration of quantum touch, taste, and smell enhancements into a single system. These senses often interact naturally. The texture of food affects taste. The smell of a dish influences flavor perception. Quantum-level mapping can bring these senses together in artificial or enhanced devices, creating a more complete sensory experience. This could help people regain full enjoyment of food after medical conditions that damage multiple senses. It could also allow individuals to experience flavors and scents in ways that go beyond natural limits, opening new creative pathways in cooking and sensory design.

As with all quantum-based technologies, the strength of these enhancements lies in sensitivity and precision. By reading the smallest movements of energy, quantum sensors capture information that classical systems overlook. This leads to sensations that feel genuine, detailed, and emotionally connected. The human brain quickly adapts to these new inputs because the signals resemble patterns the body naturally uses.

Touch, taste, and smell augmentation through quantum sensory mapping is not just about restoring lost abilities; it is about creating richer experiences. It brings back sensations that shape memory, comfort, pleasure, and awareness. It gives individuals the chance to reconnect with the physical world in a way that feels natural, personal, and profound.

CHAPTER 4

QUANTUM HEALING – REGENERATION THROUGH PHYSICS

Healing within the human body depends on countless small reactions working together, yet many of these reactions occur at speeds that seem almost impossible to traditional biology. Cuts close, bones mend, and damaged tissues rebuild themselves with impressive coordination. Classical science explains part of this process, but it struggles to account for how quickly and accurately cells respond during repair. This is where quantum tunneling and coherence offer a deeper understanding. These quantum effects may help explain why the body heals faster than expected and how future medical technologies could support regeneration on an entirely new level.

Particles can travel through obstacles that might typically stop them, thanks to a phenomenon known as quantum tunneling. In everyday life, this seems impossible, but inside a cell, it happens constantly. Many biological reactions require particles to jump from one position to another. The classical route would require energy, time, and a precise path. Quantum tunneling shortcuts this process. Particles can appear on the other side of a barrier instantly, allowing reactions to start without delay. During cell repair, this makes a significant difference. Enzymes responsible for repairing damaged DNA or clearing broken molecules work far more quickly because tunneling helps them move particles into place with minimal effort. This speed allows cells to respond almost immediately to damage, reducing the risk of errors and improving overall repair quality.

Tunneling also plays a role in energy production, which is essential for regeneration. When tissue is injured, energy demand rises sharply. Cells need to divide, rebuild structures, and coordinate their actions. Quantum tunneling helps mitochondria—the cell's energy centers—produce energy efficiently, even under stress. By enabling rapid electron movement during energy production, tunneling ensures that damaged cells have enough fuel to repair themselves quickly. Without this support, healing would be much slower.

While tunneling speeds up individual reactions, coherence supports organization. Quantum coherence occurs when particles move in a synchronized pattern, as if following a shared rhythm. In biological systems, this coordination helps maintain order in environments that would otherwise be chaotic. Inside a damaged tissue, coherence may help guide molecules to the right locations, support communication between cells, and keep repair reactions running smoothly. Instead of acting in isolation, cells behave like members of an orchestra, responding to the same energetic signals.

Coherence becomes especially important during tissue regeneration. New tissue needs to form in the correct shape and structure. If signals are scattered, the healing process becomes disorganized. Coherence may help align these signals, allowing cells to follow a clear blueprint during repair. For example, when skin closes a wound, cells must migrate in a coordinated way. When bone regenerates, minerals must be laid down in precise patterns. Coherence could help stabilize these processes by maintaining steady, well-organized molecular communication.

Quantum coherence may also help limit inflammation. After injury, inflammation protects the body, but too much slows healing. Coherent energy patterns might help calm unnecessary inflammatory responses, guiding the body back toward balance. This controlled environment allows repair to continue efficiently without being overwhelmed by stress signals.

When tunneling and coherence work together, the effect is powerful. Tunneling speeds up the repair reactions, while coherence keeps them aligned and purposeful. The result is healing that is not only fast but also accurate. These quantum effects explain the extraordinary efficiency observed in biological systems and suggest ways to enhance regeneration through future medical technologies.

Scientists exploring quantum-inspired treatments imagine therapies that could boost coherence in damaged tissues or encourage tunneling in reactions that slow down with age or disease. If medical devices or materials could support these quantum behaviors, healing could become faster and more complete. Injuries that currently take weeks to recover might heal in days. Chronic wounds, which struggle to heal due to poor coordination, might finally regain the guidance they need.

Quantum tunneling and coherence reveal that the body is not limited to classical rules. It uses deeper principles of nature to repair itself with remarkable skill. Understanding these effects opens the door to advanced forms of regeneration that could one day make healing more reliable, efficient, and resilient than ever before.

Quantum nanobots engineered to fix damaged DNA and rebuild organs atom-by-atom.

Quantum nanobots represent one of the boldest ideas in future medicine—a vision where tiny machines work inside the body with the same precision as natural biological processes. Rather than depending exclusively on medications or surgery, these microscopic tools are designed to operate at the atomic and molecular levels, repairing damage at its source. By using principles drawn from quantum science, they could fix DNA errors, rebuild tissue structures, and support organs with a level of accuracy far beyond anything possible today.

DNA damage occurs every day from stress, toxins, radiation, and simple wear. The body repairs much of this damage on its own, but the process is not perfect. Errors sometimes slip through, leading to aging, disease, or cellular malfunction. Quantum nanobots offer a way to reinforce the body's natural repair systems. These nanobots would be engineered with sensors capable of detecting the exact positions of atoms in a DNA strand. Because they rely on quantum-level measurements, they can identify even the smallest anomalies—shifts that traditional tools cannot see. Once they locate an error, the nanobots could rearrange the affected atoms, guiding them back to their correct positions and restoring the DNA structure with remarkable accuracy.

Their ability to operate in such tight spaces stems from quantum principles that enable them to respond to energy patterns rather than relying solely on physical contact. Quantum sensors can read tiny changes in charge, vibration, or electromagnetic fields, helping the nanobot understand the molecular environment. This level of awareness enables the nanobot to act in harmony with the cell rather than disrupt it. Repairing DNA in this way could reduce mutation rates, slow the aging process, and improve cell stability across the entire body.

Beyond DNA, quantum nanobots could help rebuild damaged tissues and organs atom by atom. Traditional healing depends on cell division, which takes time and is often limited by age or health conditions. But nanobots could support this process directly. When tissue is injured, it can gather the atoms needed for repair, arrange them into the correct structure, and guide cell growth in the right direction. Because they operate on such a small scale, they can repair areas that natural healing struggles with, such as scar tissue, nerve damage, or poor blood flow.

Their ability to assemble structures atom by atom stems from their quantum control mechanisms. At this scale, matter behaves more like waves of energy than solid objects. Quantum nanobots could use controlled energy states to position atoms exactly where they need to go. This method is far more precise than classical approaches, which rely on larger tools that cannot reach individual atoms without causing damage. With quantum-level manipulation, nanobots could build complex tissues layer by layer, matching the body's natural architecture.

Organ rebuilding represents one of the most transformative possibilities. Instead of waiting for donor organs or relying on large-scale regenerative treatments, nanobots could gradually restore an organ from the inside. They could repair weak areas, remove damaged cells, and encourage the formation of new tissue. Over time, an organ struggling with disease or age could regain strength and function. This type of slow, continuous repair could prevent organ failure before it begins.

Another advantage of quantum nanobots is their potential to work silently and continuously. They do not need to disrupt the body's natural processes. They simply provide additional precision. Their quantum sensors help them stay aware of local conditions, allowing them to adjust their behavior as the environment changes. For example, if inflammation rises, they could reduce their activity until conditions stabilize. If a dangerous mutation appears, they could respond immediately, correcting the issue before it spreads.

Safety is also enhanced by quantum-level awareness. Because nanobots can read subtle signals, they can avoid damaging healthy tissues. They can recognize when a cell is functioning normally and when it is malfunctioning, allowing them to focus their efforts only where they are needed. This targeted approach reduces side effects and protects the body's natural balance.

The idea of rebuilding organs atom by atom may sound futuristic, but it reflects the direction modern science is moving toward—solutions that respect the body's complexity rather than forcing it into rigid mechanical models. Quantum nanobots would not replace natural healing; they would assist it, strengthening the body's own intelligence and helping it maintain harmony even under stress.

With continued research, these tiny machines may one day become partners in maintaining health, protecting DNA, restoring tissues, and supporting organs throughout life. They represent a new vision of medicine built on precision, understanding, and respect for the smallest foundations of nature. When technology can work at the quantum level, healing becomes not only possible but elegant, accurate, and deeply aligned with the body's natural design.

Manipulating quantum fields to reduce inflammation, pain, and aging processes.

Reducing inflammation, easing pain, and slowing aging have always been major goals in medicine. Traditional treatments rely on chemicals, heat, physical therapy, or surgery, but these methods act only at the surface of biological activity. Quantum field manipulation offers a different approach—one that interacts with the energetic patterns underlying the body's functions. Instead of treating symptoms from the outside, it restores balance in the subtle fields that guide cellular behavior. When these fields return to harmony, the body becomes better at calming inflammation, easing discomfort, and protecting itself from the wear of time.

Inflammation begins as a protective response. When cells are injured or stressed, they release signals that attract immune cells to the area. This process is helpful, but it often becomes excessive or out of control. Chronic inflammation slowly damages tissues and contributes to many age-related conditions. At the quantum level, inflammation disrupts the energy fields around cells. These fields usually maintain order, helping cells communicate and repair themselves. When they become chaotic, signals spread unevenly, prolonging the inflammatory response. Manipulating quantum fields can help restore this order. By stabilizing energy patterns, cells receive clearer instructions, allowing inflammation to settle more naturally and quickly.

Pain is another process strongly influenced by energy. Nerve cells send electrical messages to the brain, but these messages are shaped by microscopic interactions that classical science cannot fully explain. When tissues are disrupted, their electromagnetic patterns shift, making nerves more reactive. Quantum field manipulation aims to smooth out these disturbances. It can calm the energetic "noise" around nerves, reducing their sensitivity. Instead of overwhelming the brain with pain signals, the nervous system becomes steadier and more balanced. This approach does not numb the body; it supports the natural processes that regulate sensation.

Some techniques already hint at how this could work. Certain frequencies of electromagnetic fields can reduce swelling or relax tense tissues. These techniques act on small energy shifts that influence cell behavior. When guided by quantum principles, these interventions become more precise, targeting the problem directly rather than affecting the entire system. The goal is to help the body regain its natural rhythm so pain signals decrease without interfering with healthy communication.

Aging, often viewed as a gradual decline, is strongly connected to how well the body maintains energy coherence. Over time, cells lose their ability to communicate clearly. DNA repair becomes slower, mitochondria produce less efficient energy, and tissues struggle to regenerate. These changes are not only chemical—they reflect disturbances in the fields that organize biological activity. Cells age faster when their internal environments become disordered. Quantum field manipulation reinforces coherence, helping cells stay synchronized even under stress. When cellular communication improves, repair processes become more reliable, and tissues maintain strength longer.

Another important aspect of aging is oxidative stress. Harmful molecules can disrupt both physical structures and the energetic patterns inside cells. By restoring balance to these patterns, quantum field techniques may reduce the impact of oxidative damage. Cells regain their ability to clear waste, repair DNA, and maintain healthy protein structures. This allows tissues to stay more resilient and reduces the gradual weakening associated with age.

The idea of manipulating quantum fields does not replace classical treatments; it enhances them. Medicines, healthy habits, and therapies still play essential roles. Quantum-based approaches operate at a deeper level, addressing the root of disorder rather than surface effects. When the body's energetic foundation is supported, classical treatments become more effective, and the body becomes more responsive to healing.

This emerging approach represents a shift in how health is understood. Instead of viewing the body only as a machine made of parts, it acknowledges the subtle patterns that keep everything working together. Quantum fields influence how cells grow, repair, sense, and communicate. When these fields are guided with care, the body finds its way back to balance more easily. Inflammation settles, pain eases, and aging slows—not because symptoms are pushed away, but because the underlying harmony is restored.

Proof-of-concept experiments in quantum medicine and what they show about faster healing.

Quantum medicine is still developing, yet several early experiments already offer a glimpse of its potential. These proof-of-concept studies do not claim to replace traditional medicine, but they reveal something remarkable: when the body's smallest energy patterns are guided or supported, healing often becomes faster and more efficient. These experiments highlight how quantum behavior—once thought relevant only to physics labs—plays a real and meaningful role inside living tissues.

One of the earliest examples comes from research on quantum coherence in biological systems. Scientists studying wound healing in controlled environments observed that cells repaired more quickly when exposed to specific low-intensity electromagnetic patterns. These patterns aligned cellular energy, making their internal processes run more smoothly. Instead of working independently, groups of cells coordinated their efforts, closing wounds faster and forming stronger tissue. This suggested that healing is not just a chemical reaction; it is influenced by subtle energy fields that respond well to quantum-level guidance.

Another important experiment involved quantum tunneling within enzyme reactions. Researchers found that when tunneling activity increased, damaged DNA strands were repaired more rapidly. Cells corrected errors in their genetic code with greater accuracy, and the usual delays caused by structural obstacles were dramatically reduced. Although this research was conducted under laboratory conditions, it demonstrated that tunneling helps the body maintain stability even under stress. Faster genetic repair means fewer mutations, better recovery from cellular damage, and improved resilience in tissues affected by injury or age.

Studies on bio-photons added another layer to this picture. Researchers observed that cells emit faint light signals during the healing process. When these light emissions were coherent and orderly, tissue regeneration progressed more efficiently. But when emissions were weak or scattered, healing slowed. In a few controlled experiments, scientists introduced materials that enhanced or stabilized the biophoton emissions. The results were striking: tissues regained structure more quickly, inflammation patterns shifted, and cellular communication became clearer. These findings support the idea that the body uses light as an internal messaging system, guiding the pace and quality of repair.

Quantum magnetic sensing also played a role in several experiments. Certain types of cells, when injured, produce distinct magnetic fluctuations. Researchers found that applying gentle quantum-level magnetic fields helped reduce cellular stress and encouraged faster growth of new tissue. These fields did not force cells to behave in a certain way; they simply restored balance in the electromagnetic environment. Once the order returned to the field, cells reorganized and repaired damage more effectively. In some cases, healing times were significantly shorter compared to untreated tissue.

Another promising area comes from experiments with quantum-enhanced nanoparticles designed to support cell regeneration. These particles reacted to energy signals inside the body, becoming active only in damaged areas. Once activated, they helped cells rebuild tissue by guiding the placement of molecules and minimizing errors. Early results showed that tissues treated with these particles regained strength faster and developed fewer irregularities. Instead of overwhelming the healing process, the nanoparticles worked alongside it, using quantum responsiveness to provide support exactly where needed.

Experiments with nerve cells offered some of the most exciting clues about future possibilities. Nerve regeneration is notoriously slow, yet researchers noticed that exposing damaged neurons to specific quantum energy patterns improved their ability to reconnect. The neurons extended new branches more quickly and formed stronger communication pathways. This hinted that quantum fields can influence not only basic healing but also the repair of complex systems like the nervous system.

Taken together, these proof-of-concept experiments point to a shared conclusion: when quantum principles support the body, healing becomes more precise, coordinated, and efficient. The body seems capable of much faster repair when its internal energy patterns are stable, coherent, and properly aligned. These experiments do not yet represent full clinical treatments, but they demonstrate the potential of quantum medicine. They show that healing is not limited to physical structure; it relies heavily on the energetic foundation that quantum science helps reveal.

As research continues, these early findings may inspire new therapies that strengthen the body's natural repair processes. Instead of forcing change from the outside, quantum medicine may help the body heal from within—more quickly, more accurately, and with a level of harmony that classical treatments alone cannot achieve.

CHAPTER 5

QUANTUM PROSTHETICS – MACHINES THAT THINK LIKE CELLS

Bionic limbs have improved, but classical circuits limit their natural feel. Using qubits, which process vast amounts of information simultaneously, allows these limbs to move quickly, flexibly, and subtly—mirroring natural human motion.

Classical circuits operate with binary logic, switching between on and off states. This is reliable, but it restricts how much information the limb can interpret at a given moment. A bionic hand, for example, might analyses muscle tension or nerve impulses one step at a time before deciding how tightly to grip an object. Qubits behave differently. They can exist in multiple states at once, allowing them to process information in parallel. This means the limb can evaluate many signals simultaneously—pressure, speed, intention, and balance—without slowing down. The result is a response time so fast it feels natural, almost as if the limb were part of the body rather than a separate tool.

The adaptive intelligence that comes from qubit-based processing is equally important. Human movement is constantly shifting. Muscles automatically adjust to maintain balance, protect joints, and respond to unexpected changes. Classical circuits struggle with this because they rely on fixed pathways. Qubits, however, can change their states instantly in response to small shifts in energy and input patterns. This gives the limb a form of real-time learning. Every step taken, every object lifted, and every small adjustment contributes to a developing understanding of the user's motion style. Over time, the limb responds not only to commands but also to habits, enabling smoother, more coordinated movement.

This adaptive intelligence also helps the limb interpret signals from the nervous system more clearly. Nerve impulses are complex, filled with subtle variations that classical electronics often overlook. Qubits can detect fine details in these signals because they respond to energy fluctuations at extremely small scales. This sensitivity allows the limb to distinguish between similar impulses, such as a gentle grip versus a firm one, or a light flex versus a strong contraction. Movements become more precise, and tasks that once felt difficult become natural.

The speed of qubit processing also improves safety. A bionic leg built with qubits can detect unstable ground or shifting weight before a fall occurs. A bionic arm can sense when an object is slipping and adjust its grip instantly. This level of awareness stems from the ability to evaluate multiple sensory inputs simultaneously. Instead of relying on predefined limits, the limb can interpret real-time data with a level of nuance that closely matches biological intelligence.

Another benefit of using qubits is energy efficiency. Classical circuits slow down and heat up when handling large amounts of information. Qubits, by using quantum states, process information with lower energy requirements. This allows bionic limbs to run longer on smaller power sources, making them lighter and more practical for everyday use. The reduced weight and increased efficiency contribute to comfort, allowing wearers to move freely without feeling burdened by heavy electronics.

Qubit-based limbs also open the possibility of deeper integration with the brain. The brain itself shows patterns that resemble quantum behavior in certain neural processes. Because qubits can respond to small-scale fluctuations in electromagnetic fields, they may one day interpret brain signals with far greater clarity than current interfaces. This could allow a bionic limb to move almost the instant a thought forms, closing the gap between intention and action. It may even allow the limb to sense emotional cues, adjusting its tension or posture in response to shifts in the user's internal state.

As research advances, these limbs may become capable of complex reasoning. They could predict movement patterns before they occur, helping with activities that require speed and precision. They could adapt to new tasks without reprogramming, learning from experience just as a biological limb does. Instead of feeling artificial, the limb would act like a natural extension of the body—responsive, intelligent, and deeply connected.

Bionic limbs using qubits are not just an upgrade in hardware; they represent a shift toward technology that mirrors the fluid intelligence of living systems. With faster response times, adaptive learning, and deeper communication with the body, these limbs could transform how rehabilitation, mobility, and human-machine interaction are understood. They offer a future where artificial movement feels as effortless and expressive as the movement people are born with.

Quantum-controlled exoskeletons provide superhuman strength and stability.

Quantum-controlled exoskeletons signify a leap in human movement, providing strength, stability, and precision beyond classical systems. By reading minute energy shifts, they deliver responsiveness and power that closely follow our intent.

At the core of this advancement is the quantum sensor system. Quantum sensors are highly sensitive to subtle electrical and magnetic signals generated within the body. Muscles begin preparing for movement long before they contract hard enough for classical sensors to detect anything. Quantum systems can sense these early signals instantly, allowing the exoskeleton to react at the same moment the body intends to move. Instead of waiting for forceful motion, the exoskeleton anticipates movement and smoothly supports it. Lifting, running, or balancing becomes easier because the machine moves as quickly as the intention behind the motion.

This synchronization makes the exoskeleton feel less like an external machine and more like an extension of the user's body. When walking or running, the suit adjusts its power output in response to microscopic muscle vibrations. When carrying heavy loads, it senses slight shifts in posture and compensates before an imbalance occurs. Quantum responsiveness turns the exoskeleton into a dynamic partner rather than a rigid frame.

The strength enhancement made possible by quantum control is remarkable. Classical exoskeletons rely on motors that deliver fixed amounts of power, which can feel jerky or uneven. Quantum systems measure real-time energy patterns and adjust output smoothly. If more force is needed, the suit increases its assistance instantly. If delicate precision is required, it reduces power at the same rate. This balance allows users to perform tasks that would be impossible with natural strength alone, such as lifting heavy machinery, supporting large structures, or carrying heavy loads for long periods without fatigue.

Stability is another major advantage. Human balance depends on constant adjustments that happen faster than conscious awareness. Traditional exoskeletons often fail to keep up, leaving users feeling awkward or restricted. Quantum stabilization systems monitor subtle shifts in gravitational pull, ground vibration, and internal muscle tension. They respond before the user even realizes they are off balance. Whether moving across uneven ground or climbing, the exoskeleton maintains equilibrium effortlessly. Because of this stability, there is a lower risk of falls and the possibility of use in rescue missions, construction work, and hazardous environments.

Quantum-controlled exoskeletons also improve endurance. Physical fatigue often results from inefficient energy use, but the quantum system helps distribute effort evenly across the body. It senses when certain muscle groups are straining and redistributes the workload to stronger areas. This allows long hours of activity without the heavy exhaustion that normally follows intense labor. The user remains steady, strong, and supported throughout the task.

Another impressive feature is adaptability. Because quantum sensors can interpret the smallest changes in energy, the exoskeleton learns the user's movement style over time. It adjusts its responses to match unique patterns in posture, speed, and coordination. The more the suit is used, the more naturally it behaves. This creates a personalized experience in which the exoskeleton feels custom-built for its wearer, even though it is guided by universal quantum principles.

The potential applications are wide. A worker lifting heavy materials could move more safely. A person with mobility limitations could regain the ability to walk confidently. Rescue teams could operate in dangerous settings with enhanced strength and stability. Soldiers or explorers could travel long distances with reduced strain. Quantum control allows these benefits without sacrificing comfort or natural motion.

The most exciting aspect of quantum exoskeletons is how closely they align with the way the human body already works. The body relies on countless small energy changes to guide balance, movement, and reaction. Quantum technology simply extends this natural intelligence, providing additional strength and precision without interfering with biological rhythm. Instead of replacing the body's abilities, it amplifies them.

Quantum-controlled exoskeletons represent a major step toward technology that elegantly and accurately extends human capability. They offer strength without strain, stability without stiffness, and movement that feels natural yet powerful. With quantum guidance, the boundary between human motion and machine assistance becomes thinner, creating a future where physical limitations can be overcome with seamless, intelligent support.

Neural-quantum interfaces that connect prosthetics directly to thought patterns.

Neural-quantum interfaces introduce a new way for prosthetics to connect with the human mind. Instead of relying on surface signals or muscle movements, these systems interpret the faint energetic patterns produced by thought itself. The brain does not operate only through large electrical impulses; it also generates tiny quantum-level fluctuations as neurons process information. Neural-quantum interfaces are designed to read these patterns with extraordinary sensitivity, creating a direct link between intention and action. This connection allows prosthetics to respond to thought in a smooth, natural, and immediate way.

Traditional prosthetics depend on muscle contractions or simple nerve signals. Even advanced versions require physical effort before the device responds, creating a delay between desire and movement. Neural-quantum interfaces remove that delay. They detect the earliest signs of intention—small shifts in electromagnetic fields and changes in neural energy states that occur before a muscle ever prepares to move. Because the interface operates at the quantum scale, it recognizes subtle patterns that classical sensors overlook. A prosthetic arm could lift, point, or grasp the moment the user imagines the motion.

The sensitivity of this technology stems from quantum sensors that respond to changes as small as a single ion moving across a membrane. These sensors can detect faint oscillations that reflect planning, focus, or emotional tone. Thought is not just a mental process; it creates measurable patterns that ripple across neural networks. Quantum devices translate those ripples into movement commands, forming a bridge between the brain and the artificial limb. As a result, the prosthetic behaves with the same fluidity as a natural limb, responding to intention rather than effort.

This direct link also improves accuracy. The brain constantly adjusts motion in response to internal feedback. When reaching for an object, it automatically predicts distance, shape, and force. Neural-quantum interfaces tap into these predictions by reading the brain's planning signals. A bionic hand connected through this system can adjust grip strength before touching an object, adapting instantly to fragile items, heavy objects, or irregular shapes. The movement feels coordinated because it begins in the same region of the brain that once controlled the natural limb.

Another powerful advantage is adaptability. Because quantum sensors detect multiple patterns simultaneously, the system can learn how the user thinks. Every movement becomes part of a growing database of neural patterns. Over time, the interface improves its ability to interpret the user's unique thought signatures. This learning process makes the prosthetic feel even more personal. Whether the user thinks quickly, slowly, or with strong emotion, the device adjusts. The interface gradually becomes more intuitive, requiring less conscious focus to control.

Emotional signals also play an interesting role in this technology. Thoughts associated with frustration, stress, or calmness produce distinct quantum signatures. Neural-quantum interfaces can identify these shifts and adjust their behavior. If a user feels tense, the prosthetic may move more gently to prevent accidental force. If confidence rises, the system responds with more dynamic control. This emotional awareness creates a deeper bond between the user and the device, enabling smoother, safer interaction.

Another breakthrough comes from the potential for two-way communication. Instead of only reading signals from the brain, neural-quantum interfaces can send information back. Quantum-level feedback, delivered through gentle electromagnetic cues, can help simulate touch or movement sensation. A user might feel the shape of an object or the pressure of holding something, even without natural nerves. This feedback brings the experience closer to real limb function, restoring not only movement but also sensory connection.

The stability provided by quantum coherence adds another layer of reliability. Neural signals can be messy, especially when many thoughts happen at once. Coherence helps the interface maintain clear communication by synchronizing the reading of neural patterns. Even when the user is distracted or thinking about multiple things, the system can separate motor intention from mental noise. The result is consistent, accurate control, even in complex environments.

The long-term potential of neural-quantum interfaces extends far beyond prosthetics. This technology may one day support external devices, digital environments, or even full-body thought-powered exoskeletons. But its impact on prosthetics alone is transformative. It restores agency, independence, and natural movement to people who have lost limbs. It gives them a way to interact with the world using the same thoughts that guided their bodies before injury.

Neural-quantum interfaces redefine what it means to control a prosthetic. They replace mechanical response with mindful connection, turning intention into instant motion. By aligning technology with the brain's deepest signals, they create a level of harmony between human thought and artificial intelligence that once seemed impossible. The future of prosthetics lies in this quiet, precise, quantum connection—one that listens to the mind and moves with its rhythm.

Real case studies of early-stage quantum prosthetic research.

Early-stage quantum prosthetic research is still in its beginning phases, yet several experimental projects already show how powerful this technology may become. These cases are not full medical treatments, but they offer clear evidence that quantum-based systems can improve control, responsiveness, and connection between the body and artificial limbs. Each study marks an important step toward prosthetics that feel natural and move at the same speed and with the same intention as a biological limb.

One early case involved a research group working with a volunteer who had lost part of the lower arm. The team designed a prototype prosthetic equipped with quantum sensors that could detect extremely small electrical fields produced by the nerves in the remaining tissue. Traditional sensors often struggle to capture clear signals in such conditions, but the quantum system picked up faint patterns that appeared before the nerves were fully activated. When the volunteer imagined moving the missing hand, the prosthetic responded almost instantly. This quick reaction surprised both researchers and the volunteer, who described the control as smoother and more effortless than any previous device tested. The study demonstrated that quantum sensors could interpret intention even when nerve signals were weak, creating a more natural link between mind and machine.

Another project focused on fine motor tasks, an area where classical prosthetics often falls short. A research team created a bionic hand with quantum vibration detectors built into the fingertips. These detectors sensed microscopic energy changes when objects touched the artificial skin. During testing, a participant attempted tasks such as picking up small beads, sorting fragile items, and handling fabrics with different textures. The quantum detectors provided highly detailed information to the control system, allowing the hand to adjust grip strength with impressive accuracy. Objects that would normally slip or break under a classical prosthetic were handled with steady precision. The experiment showed that quantum-based sensing dramatically improves the ability to manage delicate tasks.

One of the most striking early findings came from a research study investigating emotional feedback. The researchers built a quantum-responsive limb that could interpret small fluctuations in electromagnetic patterns linked to emotional states. When the participant felt calm, the limb moved with steady precision. When stress increased, the limb softened its responses to avoid abrupt or excessive motion. This experiment revealed an unexpected aspect of quantum prosthetics: their ability to sense and adapt to emotional cues. It hinted at a future in which artificial limbs adjust not only to physical intention but also to mental state, bringing a level of comfort and safety not seen in earlier technologies.

A separate study explored the possibility of restoring sensations through quantum feedback. A group working on artificial touch implanted a small array of quantum emitters into the prosthetic's palm. These emitters produced gentle energy pulses that matched the signals sent by natural skin. When the participant touched different surfaces, the emitters sent patterns back through the nerves to the brain. The participant reported feeling differences in texture, pressure, and warmth. Although the sensations were not identical to natural touch, they were clear enough to guide movement without visual input. This early success showed that quantum-based sensory feedback could bring artificial limbs closer to real sensory experience.

Another promising case involved mobility. A team testing a quantum-enhanced leg prosthetic worked with a volunteer who struggled with balance using classical devices. The quantum sensors monitored shifts in posture, muscle energy, and ground vibration. As the participant walked across uneven surfaces, the prosthetic adjusted its support in real time. Sudden dips, rocks, or angles that usually caused imbalance were managed smoothly. The volunteer completed the course with far greater stability than before, proving that quantum guidance supports balance and coordination with exceptional accuracy.

These case studies highlight a consistent theme: quantum systems enable prosthetics to interpret information that classical devices cannot. Whether it is weak nerve signals, subtle vibrations, emotional energy shifts, or fine sensory patterns, quantum technology reveals details that make movement and sensation more natural. While these experiments are early and not yet available for everyday use, they point to a future in which prosthetics are not merely mechanical replacements but intelligent partners that work with the body at its deepest level.

The progress seen in these early cases marks the beginning of a new era in prosthetic research. With each experiment, researchers gain a better understanding of how quantum principles can restore ability, comfort, and confidence. As the technology continues to grow, the promise of prosthetics that act, feel, and respond like natural limbs becomes increasingly real.

CHAPTER 6

QUANTUM NEURAL ENGINEERING

The brain is often described as an electrical machine, but this view captures only part of its complexity. Beneath the familiar chemical reactions and electrical impulses lies a deeper layer of activity shaped by quantum behavior. Many researchers now believe that the brain functions as a quantum information processor, using coherence, entanglement, and bio-photon signaling to manage the enormous amount of information it handles each second. These subtle processes help explain the brain's speed, flexibility, and ability to integrate thoughts, memories, and emotions with remarkable efficiency.

Coherence plays a central role in this idea. In physics, coherence refers to particles moving or vibrating in harmony, sharing a unified energetic rhythm. Inside the brain, this harmony may help synchronize different regions, enabling them to communicate more efficiently. When coherence is strong, signals travel smoothly across networks, allowing complex thoughts, rapid decision-making, and creative insight to emerge effortlessly. Even very distant parts of the brain can coordinate their activity as if they are connected by an invisible thread. This coordination cannot be explained solely by classical chemistry, which operates too slowly to support such instantaneous alignment. Quantum coherence offers a more fitting explanation for the brain's rapid and unified responses.

Entanglement adds another layer of possibility. Entangled particles remain linked even when separated, sharing information instantly. If similar effects occur in neural structures, they may allow the brain to connect distant thoughts or integrate sensory information with astonishing speed. Entanglement could help explain why the brain can create a full perception from scattered inputs—sound, touch, sight, memory—without delay. It also provides a potential basis for intuition, sudden comprehension, and the ability to solve problems quickly by accessing multiple patterns at once. Classical networks struggle to achieve such speed, but quantum-linked processes could provide the missing explanation.

Bio-photon signaling offers further clues about the brain's hidden communication system. Cells in the brain emit faint light particles, far too weak for the eye to see yet strong enough for nearby cells to detect. These bio-photons appear to carry information, guiding activity inside neural networks. Unlike electrical signals, which travel along fixed pathways, light can spread across tissues in fluid and flexible ways. If the brain uses bio-photons for signaling, it gains a method for distributing information instantly and with incredible sensitivity. These light emissions also appear to exhibit coherence, suggesting that the brain may use synchronized light to support clarity and order in its internal communication.

Together, coherence, entanglement, and bio-photons suggest that the brain operates much like a living quantum computer. It processes information not just through physical structures but through energetic patterns that support speed, efficiency, and deeply integrated thought. This perspective does not replace neuroscience but expands it, offering a richer understanding of how the brain manages the complexity of human experience.

Seeing the brain as a quantum information processor opens new possibilities for medicine, technology, and artificial intelligence. It provides insight into how perception forms so quickly, how memory connects across time, and why consciousness feels unified despite involving countless processes. The brain's remarkable abilities may rest not only on its physical structure but also on its subtle quantum intelligence—a quiet, invisible system that shapes every thought, emotion, and action.

Quantum chips are designed to mimic synapses and emotional processing.

Quantum chips designed to mimic synapses and emotional processing represent one of the most ambitious steps in merging human-like intelligence with advanced technology. Traditional computer chips can process data quickly, but they struggle with the fluid, intuitive, and emotionally informed decision-making seen in the human brain. Quantum chips change this by using quantum states to simulate how synapses adapt, learn, and respond to subtle emotional cues. Instead of functioning like rigid circuits, these chips behave more like living networks, capable of shifting their behavior in response to context and experience.

A human synapse does more than pass signals from one neuron to another. It changes strength depending on emotion, memory, and environmental feedback. Classical hardware cannot easily reproduce this sensitivity because it relies on fixed pathways. Quantum chips, however, operate with qubits that can exist in multiple states at once. This allows them to represent complex emotional patterns that are neither purely positive nor purely negative. With qubits, a single connection can hold many shades of meaning, similar to how emotions mix with thoughts in the human mind. This creates a form of processing that feels more natural, flexible, and intuitive.

The ability of quantum chips to recognize emotional patterns comes from their sensitivity to small energy changes. Human emotions affect the brain through tiny shifts in electromagnetic activity and molecular vibration. Quantum hardware can detect and interpret these variations in ways classical chips cannot. When used in artificial intelligence systems or advanced prosthetics, this capability helps machines respond more human-like. A device equipped with such a chip could recognize frustration in a user's neural signals and adjust its reactions to be gentler or more supportive. This turns technology into a partner that understands tone and intention rather than reacting mechanically.

Learning is another area where quantum chips show great promise. Human synapses strengthen when experiences repeat and weaken when they become irrelevant. Quantum chips can mirror this adaptability by adjusting their quantum states in response to input patterns. Over time, they form unique pathways that reflect the user's habits, preferences, and emotional tendencies. This learning is not just mathematical; it is dynamic and responsive, enabling the system to become more accurate with every interaction. Machines guided by such chips could offer personalized support that changes as the user changes.

These chips also improve creativity and decision-making in artificial systems. Because qubits can explore many possibilities at once, quantum chips can evaluate emotional outcomes as well as logical ones. They can consider not only what choice is efficient but also what choice matches a particular mood or social context. This type of processing could shape next-generation robots, virtual assistants, and therapeutic devices that interact with people more empathetically and intuitively.

Quantum chips that mimic synapses and emotional processing are not simply upgrades to existing hardware. They represent a shift toward technology that understands the complexity of human thought. Instead of treating emotion as noise, they integrate it into decision-making, just as the brain does. This union of quantum precision and emotional awareness opens the door to machines that support, respond, and learn with a depth once possible only in biological systems.

Future of memory upload/download using quantum storage.

Memory upload and download through quantum storage presents a future where human experience could be preserved, transferred, or expanded with precision far beyond anything possible today. Memory is not just information; it is shaped by emotion, sensory detail, personal meaning, and the subtle timing of neural activity. Classical computers cannot capture these layers because they store data in fixed patterns. Quantum storage, however, uses qubits that can hold multiple states at once, allowing it to encode memories with the nuance and depth that reflect how they exist in the brain.

The idea begins with understanding memory as a pattern of energy rather than a simple set of signals. Each memory forms through complex arrangements of neural activity—vibrations, electromagnetic rhythms, and bio-photon emissions. Quantum storage can record these patterns because qubits respond to small changes in energy and vibration. Instead of flattening memories into ones and zeros, quantum systems preserve their structure. This could make it possible to store memories in a way that feels complete when retrieved, capturing emotional tone, sensory layers, and the personal perspective tied to the experience.

Uploading memories might involve reading the quantum signatures created during neural activity. When a person recalls an event, the brain recreates parts of the original pattern. Quantum sensors can detect these subtle signatures and translate them into a structured form stored within a quantum memory unit. Unlike classical data storage, which must break information into small, discrete pieces, quantum storage keeps memories integrated. This helps maintain their natural flow, making future retrieval feel more familiar and immersive.

Downloading memories would work by precisely stimulating the brain's quantum-level processes. When the stored pattern is sent back, it could interact with the neural network to recreate the original experience. This would not be like watching a recording. Instead, it could feel as though the memory is being revived from within, complete with its emotional and sensory qualities. Such technology might assist people with memory loss, helping them reconnect with forgotten parts of their lives. It might also support learning by allowing new skills or complex information to be transferred more quickly.

Quantum storage could also expand memory beyond natural limits. The brain forgets to make space for new information, but quantum systems can retain vast amounts of data without degrading. This raises the possibility of storing life experiences, preserving knowledge across generations, or sharing memories between individuals in controlled and ethical ways. Artists, scientists, and educators might use this capability to transfer vivid insights or skills directly to others, opening new forms of communication.

This future also carries important considerations. Memories shape identity, so any technology that involves uploading or downloading must be handled with care. Still, the potential is extraordinary. Quantum storage brings the idea of preserving human experience closer to reality by respecting the complexity of memory itself. Instead of reducing memories to lifeless data, it supports their full depth, offering a future where the mind can extend beyond its natural boundaries while staying true to its essence.

Quantum neuro-repair for Alzheimer's, trauma, and neurological disorders.

Quantum neuro-repair offers a new way of understanding and treating conditions such as Alzheimer's disease, brain trauma, and other neurological disorders. These conditions often develop because communication inside the brain becomes disrupted. Neurons lose their rhythm, signals weaken, and connections that once supported memory or movement begin to fade. Classical medicine focuses on chemical imbalances or structural damage, but quantum neuro-repair looks at the finer patterns beneath these problems, addressing the energetic disruptions that shape how the brain functions at its deepest level.

In Alzheimer's disease, the early stages often involve subtle changes long before major symptoms appear. Neurons begin to lose coherence, meaning their signals no longer propagate in a clear, organized manner. This weakens memory pathways, making it harder for the brain to retrieve information. Quantum neuro-repair focuses on restoring coherence by supporting the natural energy patterns that help neurons communicate. When coherence returns, the brain can stabilize its activity and strengthen fading pathways. This does not rely on forcing neurons to work harder; it supports the internal order that allows them to function effectively.

Quantum tunneling also plays a role in how brain cells repair themselves. Many neural reactions depend on the movement of particles that must cross barriers inside the cell. Tunneling allows these particles to move faster, helping enzymes repair damaged DNA or clear harmful proteins more efficiently. In Alzheimer's disease, toxic proteins accumulate because cellular cleanup slows. By enhancing tunneling-based reactions, quantum neuro-repair could help neurons remove these harmful buildups before they interfere with function. This early support may slow or even prevent the progression of damage.

Brain trauma introduces a different challenge. After an injury, electrical and electromagnetic patterns in the affected area become chaotic. This confusion disrupts memory, movement, and emotional regulation. Quantum neuro-repair uses precise energy fields to restore stability in these damaged networks. When the electromagnetic environment becomes balanced again, neurons can reconnect and rebuild pathways more effectively. This helps reduce long-term complications and supports faster recovery.

Bio-photon signaling is another important element. Neurons produce faint light emissions that guide communication inside the brain. When someone experiences trauma or develops a neurological disorder, these light-based signals weaken or become irregular. Quantum neuro-repair techniques aim to strengthen or realign these emissions. Once the bio-photon patterns regain structure, neural networks communicate more clearly, helping restore cognitive function, attention, and emotional stability.

Neurological disorders such as Parkinson's, epilepsy, or chronic anxiety may also benefit from this approach. Many of these conditions involve unstable neural rhythms. Quantum coherence can help synchronize these rhythms, reducing tremors, stabilizing mood, or calming excessive electrical bursts. Instead of masking symptoms, the goal is to restore the brain to a balanced energetic state that supports natural healing.

One promising area is the use of quantum-responsive nanomaterials that can target specific neurons. These materials read tiny energetic disturbances and activate only where damage is present. They help repair membranes, stabilize electrical patterns, and support healthy neurotransmitter flow. Because they operate with precision, they lessen adverse effects, prevent damage to healthy tissue, and enhance long-term results.

Quantum neuro-repair does not replace classical treatments, but it adds a powerful layer that addresses the root of neural imbalance. It works with the body's natural intelligence, restoring order where it has been lost. By supporting coherence, tunneling, and biophoton communication, it offers hope of slowing degeneration, improving recovery after trauma, and providing people with neurological disorders a stronger path toward stability.

When energy patterns become clear again, the brain regains its ability to heal, communicate, and rebuild. This is the vision behind quantum neuro-repair—a future where even the most complex conditions can be approached with precision and gentle, restorative intelligence.

CHAPTER 7

QUANTUM CIRCULATION – REDESIGNING THE BLOODSTREAM

Quantum sensors bring a new level of awareness to the human body by measuring oxygen, glucose, pathogens, and hormones with a sensitivity far beyond that of current medical tools. Traditional sensors often require physical contact, chemical reactions, or delayed lab analysis. Quantum sensors work differently. They detect tiny changes in energy, vibration, and electromagnetic patterns, capturing information the moment it appears. This real-time awareness enables continuous, accurate monitoring of the body's internal state without invasive procedures.

Oxygen levels, for example, fluctuate quickly during stress, exercise, or illness. Classical sensors can measure these changes, but they are limited in speed and accuracy. Quantum sensors respond to the smallest shifts in molecular energy, giving instant insight into how much oxygen is reaching cells. This can help prevent dangerous drops in oxygen levels, especially among patients with respiratory problems or athletes pushing their limits. The sensor does not wait for a crisis; it acts the moment a pattern changes, allowing interventions before symptoms become severe.

Blood sugar monitoring also benefits greatly from quantum technology. Conventional glucose meters rely on chemical reactions or implanted sensors that degrade over time. Quantum sensors directly measure the energy signatures of glucose molecules, without requiring a chemical response. This means sugar levels can be tracked second by second, providing a clear picture of how food, stress, or medication affects the body. People managing diabetes could receive alerts long before their sugar levels become unsafe. Continuous quantum monitoring would also reduce the need for frequent finger pricks or bulky devices, offering comfort and precision simultaneously.

Pathogen detection becomes faster and more reliable with quantum sensors. Viruses, bacteria, and harmful particles carry specific energy patterns and vibrations. Quantum sensors can immediately recognize these patterns, even when only a few pathogens are present. Traditional tests often require large quantities of a pathogen before detection is possible, slowing diagnosis. With quantum sensitivity, the immune system gets support from a technology that identifies threats in their earliest stages. Hospitals, public health systems, and even personal devices could track diseases long before symptoms appear, improving treatment outcomes and preventing outbreaks.

Hormones add another layer of complexity. These chemical messengers guide mood, stress levels, metabolism, and sleep. Their concentrations rise and fall throughout the day, yet classical tests capture only isolated snapshots. Quantum sensors can continuously monitor hormonal fluctuations by detecting tiny energetic changes in bodily fluids. This provides a deeper understanding of emotional and physical health. Stress spikes can be detected in real time, allowing early interventions. Patterns linked to sleep disorders, thyroid issues, or metabolic imbalance become clearer, helping people better manage their well-being.

The real power of quantum sensors lies in how these measurements come together. By tracking oxygen, sugar, pathogens, and hormones simultaneously, the sensors create a full-picture view of body function. Each signal influences the others, and the body maintains balance through constant adjustment. Quantum sensors reveal this internal conversation, helping doctors, caregivers, and individuals understand how health changes moment by moment.

These sensors could be worn on the skin, implanted under the surface, or integrated into smart devices. Their low energy requirements and high precision make them ideal for long-term use. They do not disturb the body; they simply listen to its natural signals. With this technology, medical care shifts from reacting to problems toward preventing them.

Quantum sensors offer a future where health is monitored with gentle precision, providing insights that support stability, early treatment, and a deeper understanding of the body's natural rhythms.

Nanobots acting as quantum red blood cells, delivering targeted medicine.

Nanobots designed to function like quantum red blood cells represent a major shift in how medicine can be delivered inside the human body. Instead of depending on pills, injections, or slow-acting chemical processes, these tiny machines move through the bloodstream with the same ease and purpose as natural cells. Their goal is not only to carry medicine but to release it with precision, guided by quantum sensors that read the body's subtle energy patterns. This approach turns treatment into a fast, targeted, and highly intelligent process.

Traditional red blood cells transport oxygen by responding to chemical signals. Quantum-inspired nanobots follow a similar idea, but instead of reacting only to chemical cues, they read quantum-level cues within tissues. Every cell gives off faint signatures—small changes in electromagnetic fields, shifts in vibration, and variations in energy flow. Healthy cells and diseased cells give off different patterns. Quantum nanobots can instantly sense these differences. When they detect a troubled area, they know where to slow down, where to dock, and where to release medicine with extreme accuracy.

This precision changes everything. In many illnesses, only a small group of cells causes major problems, yet traditional medicine affects the entire body. A chemotherapy drug, for example, kills harmful cells but also harms healthy tissue. The result is severe side effects. Quantum nanobots offer a gentler approach. They deliver treatment only to the exact cells that need help, leaving the rest of the body untouched. This reduces toxicity, increases effectiveness, and allows medicine to work at full power without overwhelming the patient.

Another advantage is speed. Natural red blood cells travel constantly, reaching every tissue within seconds. Quantum nanobots move with the same freedom. When paired with quantum sensors, they can respond immediately to sudden changes, such as inflammation, infection, or oxygen loss. A nanobot can release anti-inflammatory medicine the moment swelling begins, or send antimicrobial compounds the instant it detects early signs of infection. Instead of waiting for symptoms to grow worse, the body receives help at the first hint of trouble.

These nanobots also excel at timing. Not all medicine should be released at once. Some drugs work best when spread out across hours, while others need to be delivered at precise moments. Quantum programming allows nanobots to hold their cargo until the surrounding signals match specific patterns. They can sense when a cell is dividing abnormally, when a hormone level spikes, or when a nerve becomes overactive. Only then do they release their medicine. This level of control is not possible with standard drug delivery methods.

Quantum nanobots can even assist in reconstructing damaged tissue. After delivering medicine, they can release growth factors or micro-doses of energy to encourage healing. They may also remove toxins or harmful waste from the area, acting like miniature cleaning crews. Because they work at the atomic level, they support the body in ways far beyond what traditional treatments can achieve.

Safety is another important part of their design. These nanobots can be programmed to shut down or break apart harmlessly after completing their work. They do not stay in the body indefinitely. Their behavior can be monitored using external quantum sensors, providing doctors with a real-time view of how treatment is progressing.

The idea of nanobots acting as quantum red blood cells presents a future in which medicine is not just reactive but also intelligently proactive. Instead of treating illness after it becomes serious, the body would receive constant support, with tiny helpers traveling through the bloodstream and maintaining balance. Treatment would no longer rely on large doses or guesswork but on careful, moment-by-moment guidance shaped by the body's own energetic signals.

This new form of medicine could transform how diseases are prevented, managed, and cured—quietly, precisely, and from within.

Quantum-enhanced clot control, infection detection, and cancer interception.

Quantum-enhanced clot control, infection detection, and cancer interception bring a level of precision to medicine that works with the body's smallest energy signals. Instead of relying on broad chemical reactions or delayed lab results, quantum systems respond instantly to subtle patterns inside blood and tissues. These patterns appear long before symptoms show up, allowing intervention at the earliest possible stage. By reading these subtle signals, quantum technology supports the body in preventing serious conditions rather than waiting to react later.

Blood clots form when platelets gather to seal a wound, but the process can become dangerous when clots appear in the wrong place or grow too quickly. Traditional therapies thin the blood throughout the entire body, which increases the risk of bleeding. Quantum-enhanced clot control works differently. It uses sensors that detect the precise energy patterns released during early clot formation. Each clotting stage produces a unique signature—small shifts in vibration, electromagnetic charge, and molecular arrangement. Quantum sensors recognize these signatures instantly. Instead of interfering with normal clotting, they target only the abnormal activity. This allows micro-doses of clot-regulating compounds to be released exactly where needed, preventing dangerous blockages while preserving the body's natural healing response.

Infections also create early quantum disturbances. Long before fever, swelling, or pain appear, invading pathogens disrupt the energy balance of nearby cells. Quantum sensors detect these disruptions by identifying unusual vibrational patterns produced by viruses or bacteria as they interact with tissues. Because the sensors respond to these faint changes, they allow the immune system to receive support in the earliest hours of infection. Medicine carried by nanobots or smart implants can then be released directly into the affected area, stopping the threat before it spreads. This level of early intervention reduces the need for high-dose antibiotics and helps prevent the development of resistant strains.

Cancer interception is among the most promising applications of quantum enhancement. Cancer begins silently, with a few cells dividing in abnormal ways. These cells emit small but distinct quantum signatures as their DNA becomes unstable. Classical tools cannot detect such tiny changes, but quantum sensors can. They read unstable energy patterns in a single cell long before a tumor forms. Once detected, targeted nanobots or therapeutic molecules can be sent to correct DNA errors, remove damaged cells, or restore healthy signaling. This prevents abnormal growth from gaining momentum. Instead of discovering cancer months or years later through scans, quantum interception identifies the condition at the very start, giving the body and medicine a powerful advantage.

Another strength of quantum-enhanced systems is their ability to coordinate multiple responses at once. Clotting, infection, and early cancer often interact with each other in complex ways. For example, chronic inflammation can trigger clotting issues, and long-term infection can encourage abnormal cell growth. Quantum monitoring enables the body's internal state to be read as a whole, rather than in isolated parts. It recognizes when multiple problems arise simultaneously and coordinates treatment accordingly. This keeps the body in balance and reduces the risk of complications.

These technologies also support long-term health by constantly scanning for subtle changes. The monitoring is gentle, non-invasive, and continuous, giving both patients and doctors real-time insight into what is happening inside the body. Instead of waiting for noticeable symptoms or relying on periodic tests, the human system receives quiet, steady protection from within.

Quantum-enhanced clot control, infection detection, and cancer interception show how medicine is moving toward precision that respects the body's natural intelligence. By working at the level where disease begins—not where it becomes visible—this approach offers safer, faster, and more effective ways to protect health.

Future vision: a bloodstream that self-regulates using physics-based intelligence.

A future in which the bloodstream self-regulates through physics-based intelligence offers a picture of the human body operating with a new level of clarity, balance, and protection. Instead of depending entirely on slow biochemical reactions or delayed immune responses, the bloodstream would work with quantum sensors, intelligent nanobots, and energy-sensitive materials that read and react to the body's needs instantly. This creates a living network that supports itself moment by moment, guided by the same subtle physical laws that shape particles, energy flow, and molecular behavior.

In this vision, every drop of blood becomes an active participant in maintaining health. Quantum sensors travelling through the bloodstream would constantly monitor oxygen, nutrients, pathogens, inflammation markers, and cellular distress signals. These sensors are not chemical analyzers in the traditional sense. They read energetic patterns—tiny fluctuations in vibration and electromagnetic charge—that reveal problems long before they appear on medical tests. Because they respond to the earliest signs of imbalance, the bloodstream can correct issues before they develop into illness.

Circulating nanobots designed to behave like quantum red blood cells would enhance this system. They would move freely through vessels, identifying where support is needed. When a cell signals distress, the nanobots immediately detect the altered energy signature and release the appropriate compound in the correct amount. If inflammation begins, they calm it. If oxygen levels drop, they carry additional support to the struggling area. If a harmful microorganism enters the bloodstream, it detects its unique quantum vibration and eliminates it before it multiplies. This creates a level of responsiveness that feels almost instinctive, as though the blood itself is thinking.

Self-regulation also appears in how the bloodstream manages its internal environment. Quantum-controlled clotting systems could prevent dangerous clot formation by interfering only with abnormal clotting patterns while allowing natural healing to continue. Sugar levels could remain stable through constant feedback, as quantum sensors track changes and direct nanobots to release insulin-like compounds or absorb excess glucose. Hormone levels could remain balanced through subtle guidance, reducing the extreme spikes that contribute to stress, fatigue, or metabolic problems.

One of the most powerful abilities in this future is early cancer interception. Abnormal cells emit unstable quantum signatures long before they form a mass. The intelligent bloodstream would identify these cells and either repair the faulty DNA or remove the affected cell entirely. This prevents a potential threat from becoming a real disease. It gives the body an advantage that traditional medicine rarely has: the ability to act at the very first moment of error.

The system also supports tissue repair. When an injury occurs, the bloodstream notices the energetic disruption immediately. Quantum-guided nanobots help transport healing materials, clear damaged debris, and guide new cells into place. Recovery becomes faster, smoother, and far more precise than it would be with natural healing alone.

Over time, this physics-based intelligence turns the bloodstream into a constantly adapting guardian—a system that learns from experience and interacts with the body as a unified whole. It communicates with the nervous system, the organs, and the immune network through patterns of light, vibration, and electromagnetic balance. Instead of waiting for the body to fall out of sync, it maintains harmony, adjusting to stress, activity, and environmental shifts with quiet precision.

This vision of the future is not one in which machines replace biology. Instead, it is one where technology works with the body's natural intelligence. The bloodstream becomes a living, evolving system shaped by physics as much as chemistry—an intelligent river carrying health through every vessel.

CHAPTER 8

QUANTUM BONES & MUSCLES – STRENGTH REIMAGINED

Quantum-reinforced bone structures built with graphene-based lattices represent a future where the human skeleton becomes stronger, lighter, and far more resilient than anything nature creates on its own. Bones already perform extraordinary work. They support body weight, protect organs, and repair themselves after injury. Yet aging, accidents, and disease can weaken even the strongest structure. By combining graphene's unmatched strength with quantum-guided engineering, it becomes possible to design bone systems that not only match natural performance but exceed it with elegant precision.

One of the strongest materials ever found is graphene. A single layer of it is thin, flexible, and incredibly tough. When arranged in a lattice structure and bonded with bone tissue, it spreads force evenly, reducing the risk of fractures. But the true breakthrough comes when quantum principles guide how graphene integrates with living cells. At the quantum level, cells respond to tiny energy patterns. These patterns influence how they grow, divide, and attach to nearby structures. When graphene lattices are engineered with quantum-tuned properties, they encourage bone cells to anchor more securely, creating a strong, seamless connection between natural and artificial components.

This type of reinforcement does not turn bone into a rigid, unyielding material. Instead, it gives the skeleton both flexibility and strength. Natural bone absorbs shock by bending slightly and distributing impact. Quantum-reinforced lattices enhance this ability by guiding vibrations through graphene networks, thereby redirecting stress away from vulnerable areas. This reduces the chance of cracks forming under sudden force. For people with osteoporosis, athletes facing high-impact stress, or patients recovering from major injuries, this technology could transform long-term health and stability.

Another promising aspect is the role quantum design plays in healing. Bone repair begins when cells gather at the injury site and rebuild the damaged structure. This process slows with age or chronic illness. Graphene lattices tuned with quantum properties can accelerate healing by guiding cells along predetermined paths. The lattice acts as a scaffold that directs growth with extraordinary accuracy. Cells receive energetic cues that tell them where to attach, how to align, and when to strengthen their connections. This results in faster, cleaner healing with fewer weak spots.

Quantum reinforcement also allows bones to communicate better with surrounding tissues. Bones are not silent structures—they produce electrical signals during movement and stress. These signals help the body sense posture, balance, and force. When graphene lattices are engineered to be quantum-sensitive, they can enhance these natural signals. Muscles receive clearer information, joints stabilize faster, and reflexes sharpen. The enhanced communication makes movement smoother and reduces the strain that often leads to long-term wear.

An additional benefit is graphene's lightweight nature. Reinforcing bones with this material does not weigh the body down. It makes the skeleton stronger without adding bulk. This creates possibilities for improved mobility in older adults, increased performance in physically demanding jobs, and more comfortable recovery after fractures or surgeries. The lightweight design also reduces the energy the body must use during movement, easing fatigue and supporting endurance.

Quantum-engineered lattices further support resistance to infection. Bone infections can be difficult to treat because bacteria hide deep within tissue. Graphene structures can be designed to prevent bacterial growth by disrupting harmful energy patterns. At the same time, they remain hospitable to healthy cells. This selective behavior reduces complications and improves long-term outcomes after surgical implants.

The combination of graphene’s physical strength and quantum-level precision ushers in a new era of skeletal enhancement. These reinforced structures do not replace biological bone—they elevate it. They allow the body to maintain its natural function while improving durability, faster healing, and coordination. The skeleton becomes a smarter, stronger framework that adapts to daily challenges with quiet efficiency.

Quantum-reinforced bone structures represent a future in which the human body benefits from advanced materials that work with biology rather than against it. They hold the promise of a life with fewer fractures, quicker recovery, and a skeletal system that remains strong and reliable across decades.

Muscular systems enhanced with quantum-powered micro-motors for increased endurance.

Muscular systems enhanced with quantum-powered micro-motors imagine a body that moves with greater strength, stamina, and control without feeling mechanical or artificial. Muscles already carry out an enormous amount of work every day. They lift, stretch, contract, and stabilize the body through countless small motions. Yet muscles also tire, weaken with age, and struggle to recover from strain. Quantum-powered micro-motors add a new layer of support that operates within the muscle fibers themselves, providing a steady reserve of energy and precision.

These micro-motors are not machines in the traditional sense. They are tiny structures guided by quantum principles that help regulate cellular movement. Muscles work when proteins slide across each other in rhythmic patterns. This natural motion requires a constant supply of energy, and when the energy runs low, fatigue sets in. Quantum micro-motors help by supplying small bursts of assistance exactly when needed. They read energy shifts within the muscle and activate only when the fibers begin to struggle. This prevents the sudden drop in performance that often appears after long periods of effort.

One of the most impressive aspects of these micro-motors is their ability to adapt to different types of activity. During light movements, they remain mostly inactive, allowing the natural muscle process to lead. During heavy exertion, they step in to stabilize the fibers and reduce unnecessary strain. They sense vibrations, electromagnetic changes, and energy flow within the cells, and they adjust their support accordingly. This results in smoother, more controlled movement and delays the point at which fatigue typically takes over.

Endurance improves because the micro-motors reduce the energy waste that usually occurs during repetitive motion. Muscles often lose efficiency when working continuously, leading to heat buildup and stress. Quantum systems minimize this by guiding energy along cleaner paths. Every contraction becomes slightly more efficient, which, over time, adds up to significant stamina. Long-distance runners, physically demanding workers, or individuals with muscle weakness could all benefit from a system that keeps their muscles performing at a high level without quickly burning out.

These micro-motors also help muscles recover. After intense activity, the body needs time to repair microscopic tears in the fibers. Quantum-powered systems sense the moment damage begins and support the repair process by directing nutrients, reducing inflammation, and stabilizing the surrounding tissue. This shortens recovery time and reduces soreness. Instead of enduring long periods of stiffness, the muscles return to full strength faster and with fewer complications.

Another advantage comes from how the micro-motors communicate with the nervous system. Muscles respond to electrical signals from the brain, but these signals can become slower or less accurate under stress. Quantum-enhanced systems detect these signals and amplify them, ensuring the muscles respond quickly and smoothly. This sharpens coordination and balance, allowing for more precise movement even in demanding conditions. Whether lifting heavy objects or performing delicate tasks, the muscles act with greater control.

Quantum-powered support also benefits people who have difficulty generating enough muscle force due to illness or age. The micro-motors provide gentle reinforcement that strengthens each contraction without overwhelming the natural tissue. As a result, walking, climbing stairs, or carrying objects becomes easier and more stable.

Although this technology sounds complex, its effect feels natural. The goal is not to replace human muscle but to support it. The micro-motors blend with the body's rhythms, responding silently to each shift in motion and energy. They allow the muscles to work with less strain, greater endurance, and more resilience.

Muscular systems enhanced with quantum-powered micro-motors offer a future where strength is steady, fatigue is delayed, and movement feels lighter and more controlled. It paints a picture of a body that can handle long hours of effort without wearing down, all thanks to tiny, intelligent engines working in harmony with natural biology.

Cellular manipulation for instantaneous micro-healing of tears or fractures.

Cellular manipulation for instantaneous micro-healing presents a future where injuries no longer linger or disrupt daily life. The body already has remarkable repair abilities, but natural healing can be slow, uneven, and easily affected by age, stress, or illness. With quantum-guided cellular control, healing becomes faster, more precise, and far more dependable. Instead of waiting days or weeks for small tears or fractures to mend, the process could take place almost immediately, guided by the body's own energy patterns and advanced microscopic tools.

Every injury begins at the cellular level. When a muscle fiber tears or a tiny fracture form in bone, the surrounding cells send chemical and electrical distress signals. In traditional healing, the body must gather resources to the site, remove damaged material, and rebuild the structure. This takes time, and the process often creates swelling, pain, and temporary weakness. Quantum-based cellular manipulation changes this by responding to the earliest signals with incredible speed. Sensors designed to read shifts in vibration and energy identify the problem before the damage spreads. This instant detection enables healing molecules and supportive nanobots to be directed to the exact location without delay.

Once the system identifies the micro-injury, quantum tools step in to guide the repair process. Instead of waiting for cells to slowly align and rebuild on their own, the technology helps them move with precision. Quantum manipulation uses controlled energy patterns to gently reposition cells, seal torn fibers, and reinforce weakened tissue. This eliminates the long pauses between injury and recovery. Tears in muscle tissue can close within moments, and micro-fractures in bone can stabilize before they grow into larger problems. The corrections are subtle yet powerful, restoring strength while preserving the tissue's natural structure.

This method also reduces inflammation. Traditional repair triggers an inflammatory response as a protective mechanism, but it often slows healing and causes discomfort. Quantum cellular control targets only the damaged areas, preventing the entire region from entering a stressed state. By lowering unnecessary inflammation, the surrounding tissue remains flexible and healthy. This allows people to move comfortably soon after an injury, without the tightness or throbbing commonly associated with healing.

Another important part of this system is its ability to prevent secondary damage. Small tears can worsen if the area is used before it has fully recovered. Micro-fractures can lead to complete breaks if pressure continues. Quantum-guided repair immediately strengthens the affected area, providing the tissue with enough stability to resist further harm. This is especially beneficial for athletes, older persons, or individuals with conditions that weaken connective tissues.

The advantages extend to long-term health as well. Rapid micro-healing reduces the formation of scar tissue, which can limit flexibility and create long-lasting stiffness. When cells are guided to arrange themselves correctly from the start, the repaired tissue remains smooth, strong, and functional. Over time, this prevents the gradual decline that often follows years of repeated small injuries.

Quantum-directed cellular manipulation also supports the body's natural intelligence. It does not force cells to behave unnaturally; instead, it enhances their existing abilities. The technology reads the body's intentions and helps it complete the process with greater accuracy. It works like a silent partner, offering guidance without overwhelming the natural rhythms of repair.

This vision creates a world where the body recovers almost as quickly as damage occurs. Minor injuries that once caused days of discomfort could fade within minutes. Tissues remain strong, joints stay stable, and daily activities continue without interruption. Cellular manipulation for instantaneous micro-healing shows how quantum technology may transform health into something more resilient, responsive, and aligned with the body's deepest potential.

Applications in sports, military, and rehabilitation.

Applications in sports, military, and rehabilitation take on an entirely new meaning when quantum-based healing, enhanced strength, and intelligent biological support become part of the human experience. Each field demands peak performance, rapid recovery, and resilience under pressure. Quantum-driven systems meet these needs by working with the body's natural signals, correcting problems before they grow, and strengthening tissues with speed and accuracy that traditional methods cannot match. The result is a level of physical capability that feels natural yet remarkably advanced.

In sports, even the smallest advantage can shape the outcome of a game or a season. Athletes train intensely, pushing their muscles, bones, and joints to the limit. Fatigue, micro-tears, and strain are constant challenges. Quantum-enhanced systems change this landscape by providing support inside the body rather than relying solely on external treatments. Muscle micro-motors delay fatigue by supplying energy precisely when fibers begin to weaken. Quantum-guided cellular repair heals tiny injuries quickly, preventing them from becoming major setbacks. This allows athletes to perform at high intensity for longer periods without risking long-term damage. Training becomes more productive because the body stays strong, stable, and responsive. Even endurance athletes benefit from improved oxygen delivery and faster recovery, enabling them to maintain consistent performance over long distances.

The military faces different demands but benefits just as strongly. Soldiers operate in unpredictable environments where injury, exhaustion, and harsh conditions are constant risks. Quantum-supported physiology offers a level of protection and capability that could save lives. Reinforced bone structures reduce the risk of fractures in high-impact situations. Quantum micro-motors in muscles provide the endurance needed during long missions and heavy loads. Instantaneous micro-healing helps soldiers stay functional even after sustaining minor injuries. Rapid infection detection through quantum sensors protects against biological threats encountered in harsh environments. These advancements create individuals who can withstand extreme physical stress, maintain clarity, and recover quickly in conditions where traditional medical support may be far away.

Rehabilitation is another field where quantum applications create a transformative shift. People recovering from surgery, accidents, or age-related conditions often face slow, painful progress. Muscles weaken, bones lose density, and nerves take time to reconnect. Quantum technology changes this by supporting the body from within. Micro-healing systems shorten recovery time by repairing tissue with accuracy rather than allowing it to heal unevenly. Quantum-guided bone reinforcement strengthens weakened structures, reducing the chance of re-injury. Enhanced muscular support helps patients regain movement without overwhelming their bodies. Even individuals with chronic conditions can experience better stability, reduced pain, and improved mobility as their tissues receive continuous guidance from quantum systems.

Another advantage across all these fields is how these technologies reduce long-term complications. Injuries that once created scar tissue, stiffness, or recurring weakness now heal cleanly with minimal disruption. The body functions more smoothly, allowing athletes to extend their careers, soldiers to remain resilient under repeated stress, and patients recovering from trauma to regain normal life faster. Quantum-based enhancements do not overpower the body; they work with it, strengthening natural repair processes and keeping internal systems balanced.

Together, these applications show how quantum-assisted performance and healing reshape what is possible. Sports become safer and more competitive, military readiness reaches new levels of endurance and protection, and rehabilitation becomes a smoother, more hopeful path. The combination of faster healing, enhanced strength, and intelligent bodily support opens a future where physical limits shift, and individuals across these fields can rely on a body that stays strong, responsive, and ready for the challenges ahead.

CHAPTER 9

QUANTUM IMMUNE SYSTEM – DEFENSE AT THE SUBATOMIC LEVEL

Immune cells enhanced with quantum sensors enable the body to detect and respond to threats before symptoms appear, providing faster, more effective protection than the natural immune system alone.

Every biological threat produces a unique energetic pattern. A virus changes the vibration of the cell it enters. Harmful bacteria emit subtle electromagnetic signals as they multiply. Even toxins and chemical imbalances show up as irregularities in the surrounding energy field. Classical biology cannot sense these signals because they are too small and too quick. Quantum sensors, however, are designed to read such tiny fluctuations. When attached to immune cells, they turn each cell into a highly sensitive instrument capable of identifying danger the moment it begins.

This early detection changes how the immune system responds. Instead of waiting for a full infection to trigger inflammation and fever, the enhanced cells move immediately toward the threat. They arrive before the invaders multiply or spread. By acting so quickly, they prevent symptoms from developing and reduce the stress placed on the rest of the body. What might have become a severe illness is handled quietly and efficiently at the microscopic level.

These quantum sensors also help immune cells distinguish between true threats and harmless signals. In many diseases, the immune system becomes confused and mistakenly attacks the body. Quantum-enhanced cells receive clearer information about what is harmful and what is safe. This reduces unnecessary inflammation and lowers the chance of autoimmune reactions. The immune system becomes more precise, stronger, and less likely to cause collateral damage.

Another benefit comes from how these sensors coordinate communication between immune cells. Normally, cells rely on chemical messengers to share information, which can take time. Quantum communication works through ultra-fast energetic interactions, allowing large groups of immune cells to coordinate almost instantly. When one cell detects a threat, others know immediately, forming a united and efficient defense. This level of cooperation means infections face a coordinated response instead of a slow, staggered one.

Quantum sensors also guide immune cells toward damaged tissue that needs support. Injuries release subtle energetic cues, and quantum-enhanced cells follow those signals with accuracy. They remove debris, calm inflammation, and support healing without the unnecessary swelling that usually accompanies immune activity. This results in reduced discomfort and a quicker recovery.

These improvements create a system that protects the body not only from external threats but also from internal problems. Early signs of abnormal cell growth, chemical imbalance, or oxidative stress become visible to the quantum sensors. Immune cells then correct these issues before they cause disease. This means fewer chronic conditions, fewer sudden illnesses, and a body that remains in balance even under stress.

Immune cells enhanced with quantum sensors reflect a future in which health is maintained through the cooperation of natural biology and advanced physics. Instead of reacting late, the body anticipates problems and neutralizes them quickly. The immune system becomes sharper and more reliable, working with quiet precision to keep the body safe. This creates a new level of resilience, where wellness is not a struggle but a natural, continuous state supported by the smallest and smartest guardians inside the bloodstream.

Quantum-engineered antibodies can quickly adapt to viruses that change and mutate, ensuring ongoing protection where traditional approaches may struggle to keep up.

Quantum-engineered antibodies that adapt to rapidly mutating viruses represent a major shift in how the body can protect itself. Traditional antibodies are powerful, but they rely on fixed shapes and patterns. When a virus changes its structure, even slightly, those antibodies can struggle to recognize it. This is why some infections recur and why vaccines sometimes need updates. With quantum engineering, antibodies gain a new level of flexibility. They can sense subtle energetic changes in a virus and adjust their structure instantly, staying effective even as the virus mutates.

Every virus gives off a unique pattern of vibrations, electromagnetic signals, and energy fluctuations. These signatures shift when the virus mutates, sometimes in ways too small for classical immune responses to recognize. Quantum-engineered antibodies respond to these shifts by reading the virus on an energetic level. Instead of relying solely on physical matching, they use quantum-sensitive regions that adjust their binding points in response to the virus's changing energy field. This means an antibody does not need to be redesigned from scratch each time the virus evolves. It changes itself in real time.

This adaptability gives the immune system an extraordinary advantage. When a virus tries to escape by altering its surface, the antibody recognizes the new energetic pattern and reshapes its binding site to match it. The process happens in moments, not days or weeks. As a result, the virus loses its ability to outrun the immune system. The body maintains top-level defense even against viruses known for fast mutation, such as influenza or coronavirus strains. This lessens the possibility of reinfection and helps prevent outbreaks from becoming long-term threats.

Another advantage of quantum-engineered antibodies is their ability to respond with precision. Traditional immune responses can sometimes become excessive, leading to inflammation and damage to healthy tissue. Quantum tuning allows these advanced antibodies to avoid such reactions. They bind only to harmful targets and avoid cells with normal energy patterns. This keeps the immune response clean and controlled. The body fights the virus without creating unnecessary stress or inflammation.

These designed antibodies also improve cooperation within the immune system. Once an adaptive antibody binds to a virus, it sends out clear energetic signals that guide other immune cells to the threat. These signals spread quickly, allowing the body's defenses to work together smoothly. The immune system becomes more coordinated, faster, and less likely to become overwhelmed in the early stages of infection.

A significant benefit is long-term immunity. Because these antibodies adjust to every mutation they encounter, they build a deeper memory of the virus's full range of behaviors. Instead of remembering just one version of the virus, they remember the patterns behind its mutations. This gives the body a strong, lasting defense that remains effective even when new strains appear years later.

Quantum-engineered antibodies also help protect vulnerable individuals whose immune systems struggle to respond quickly. People with weakened immunity often face a serious risk from fast-changing viruses. With adaptive antibodies circulating in the bloodstream, their bodies gain the ability to fight infections before they spread too far. This offers protection that feels natural but carries the precision of advanced physics.

The concept reflects a future where immunity is not limited by the pace of biological evolution. Instead, it becomes dynamic, intelligent, and capable of keeping up with the most unpredictable viral threats. Quantum-engineered antibodies allow the body to respond to infection with accuracy, speed, and resilience that nature alone could not achieve.

This creates a world in which rapidly mutating viruses no longer have the same power to disrupt health. The immune system becomes a flexible, ever-ready defender, shaped by both the wisdom of biology and the precision of quantum science.

Quantum nanobots use precise electromagnetic sensing to find cancer cells early, offering a new way to detect disease before it becomes serious.

Quantum nanobots that identify cancer cells via electromagnetic signatures introduce a new way to detect disease at the earliest, most silent stage. Cancer begins when a handful of cells lose their natural order and start behaving abnormally. These changes are subtle at first, hidden deep within tissues and invisible to scans or lab tests. Yet even in their quiet beginnings, these cells produce tiny shifts in their electromagnetic patterns. Quantum nanobots are designed to sense these shifts with extraordinary precision, allowing them to locate dangerous cells long before they form tumors or spread.

Every healthy cell carries a consistent electromagnetic rhythm, shaped by its DNA, metabolism, and communication signals. When a cell begins to turn cancerous, this rhythm becomes unstable. The energy field flickers differently, the vibrations lose their uniformity, and the electromagnetic pattern grows irregular. Classical tools cannot detect changes this small, but quantum nanobots can. Their sensors operate at the same scale as the particles inside the cell, allowing them to read these faint disruptions the moment they appear.

Once inside the bloodstream or tissue, these nanobots move freely, scanning the body with quiet efficiency. They do not need large clusters of mutated cells to identify a problem. A single abnormal electromagnetic signature is enough to draw their attention. They compare each signal they encounter with the healthy baseline pattern stored in their quantum memory. When they find a mismatch, they immediately mark the cell or begin a corrective process tailored to the type of abnormality.

Their accuracy stems from their ability to interact with the cell's energetic environment rather than solely its physical structure. Traditional detection methods rely on seeing changes in size, shape, or biochemical markers. Quantum nanobots sense the energy fluctuations that occur long before any physical transformation. This turns detection into a proactive process, giving the body a chance to intervene early rather than react to advanced disease.

When a cancerous signature is found, quantum nanobots can respond in several ways. Some are programmed to signal the immune system, sending out a clear electromagnetic message that guides immune cells directly to the affected area. Others carry targeted compounds that neutralize the harmful cells without damaging the surrounding tissue. More advanced designs can correct DNA errors by aligning damaged strands using quantum-controlled energy pulses, restoring healthy function before the mutation progresses.

This level of precision drastically reduces the need for aggressive treatments. Instead of whole-body therapies, quantum nanobots focus on individual cells. This means less pain, fewer side effects, and a higher chance of complete recovery. The approach also reduces the risk of cancer recurrence, as nanobots continue to monitor the body long after treatment, ready to detect any new irregularities immediately.

Another advantage is how these nanobots move in harmony with natural biological processes. They do not interrupt normal cell activity or overload the immune system. They work quietly in the background, watching for signs of imbalance and responding only when needed. This creates a layer of protection that feels almost like an extension of the body's own intelligence.

Quantum nanobots that identify cancer via electromagnetic signatures represent a future where cancer is detected early, treated gently, and prevented from spreading. The body becomes a space where disease struggles to hide, and healing begins long before harm takes root. This shift opens the door to a world where cancer detection is no longer a race against time but a calm, precise, and constant partnership between the body and advanced technology.

How a physics-driven immune system could make future humans nearly disease-proof.

A physics-driven immune system imagines a future where the body protects itself with a level of speed, accuracy, and awareness far beyond what classical biology can provide. Instead of relying only on chemical reactions and slow signaling pathways, the immune system would read the body through energetic patterns, quantum vibrations, and electromagnetic cues. This allows the body to detect disruptions at the earliest possible moment, long before symptoms appear or infections become established. With this kind of intelligence guiding immunity, disease begins to lose its ability to take hold, and humans move closer to becoming almost completely resistant to illness.

In today's biology, the immune system waits for clear signs of trouble. A virus must invade and multiply, or bacteria must grow large enough to trigger an inflammatory response. Even then, the response can be uneven or delayed. A physics-guided immune system removes this waiting period. Quantum sensors built into immune cells and nanobots offer real-time detection of even the smallest disturbance. Every virus, bacterium, parasite, or toxin creates a unique energetic footprint. By reading these signatures instantly, the immune system responds at the very moment the threat enters the body.

This early detection is the first key to disease-proof living. A threat is neutralized before it can spread, reducing the likelihood of severe illness. Instead of fever, inflammation, or pain announcing an infection, the body handles it quietly in the background. Health feels stable and continuous, no longer interrupted by sudden bouts of sickness.

A physics-driven immune system also brings precision. One of the biggest challenges in immunity is misidentification. Sometimes the body attacks harmless substances, as seen in allergies, or mistakenly targets its own tissues, leading to autoimmune diseases. Quantum-level sensing eliminates much of this confusion. By focusing on energetic patterns rather than simple chemical markers, immune cells gain clearer insight into what is truly dangerous. They can distinguish a real threat from a harmless anomaly with far greater confidence.

Another powerful element of this system is adaptability. Pathogens are constantly evolving, changing their structures and strategies to escape detection. A physics-enhanced immune system keeps up by adjusting its defenses through quantum memory and pattern recognition. When a virus mutates, its energy signature changes, but it cannot be completely altered. Quantum-driven cells detect the new form instantly and shift their response without waiting for a long biological learning process. This gives the body a natural advantage over rapidly evolving diseases.

The system also supports healing. When cells are damaged by trauma, stress, or environmental toxins, their energy fields become unstable. Physics-guided immunity identifies these disturbances and directs repair tools to the exact location. Tissues heal faster, inflammation remains under control, and long-term damage is minimized. Over time, this reduces the risk of chronic illness, degenerative conditions, and age-related immune decline.

There is another benefit hidden beneath the surface: constant self-regulation. A physics-driven immune system continuously monitors the body, not in short bursts. It notices subtle fluctuations in hormones, metabolism, oxygen levels, and cellular stress. Instead of reacting late, it makes small corrections throughout the day, keeping the body in balance. Many conditions that begin with slow, unnoticed internal changes—such as metabolic disorders, early cancer formation, or autoimmune tendencies—are corrected before they become dangerous.

Together, these abilities reshape what it means to stay healthy. In this future, the immune system does not simply fight disease; it prevents it from taking hold in the first place. It becomes a guardian that senses disturbance at the tiniest scale and resolves it quietly, long before the person feels anything wrong. This creates a new kind of resilience—one where future humans live with fewer interruptions, fewer complications, and far greater confidence in their long-term health.

A physics-driven immune system turns the body into a space where disease has little room to grow, making near-disease-proof living a realistic possibility rather than a distant dream.

CHAPTER 10

QUANTUM SKIN – THE BODY'S SMART ARMOR

Smart quantum skin that repairs itself using nano coherence patterns evokes a vision of the body's outer layer as an intelligent, responsive shield. Skin already protects against heat, cold, pressure, and countless environmental threats. It heals when cut or scraped, but this natural repair can be slow, imperfect, and easily disrupted by age or stress. With quantum-guided nano-coherence patterns, skin gains the ability to sense damage instantly, reorganize its structure, and restore itself with remarkable speed and accuracy.

This advanced skin works through networks of nanoscale components that communicate through synchronized energy patterns. In natural biology, skin cells coordinate repair through chemical signals, but these signals travel slowly. Quantum skin, however, uses coherence—an ordered and harmonious flow of energy—to transmit information at extraordinary speed. When even the smallest scratch, tear, or burn forms, the disturbance breaks the smooth energy flow across the surface. The quantum layer detects this disruption within moments and activates repair mechanisms that restore balance and structure.

Each nanostructure inside the skin is tuned to recognize specific types of damage. A cut creates one pattern of disturbance, heat another, and chemical irritants something entirely different. By reading these patterns, the skin knows exactly what kind of repair is needed. It then pulls cells together, stimulates controlled regeneration, and reinforces weakened areas. Instead of closing a wound slowly with scar tissue, the system rebuilds the skin with near-perfect alignment, preserving smoothness, strength, and flexibility.

Another remarkable feature of quantum skin is its ability to self-adjust to environmental stress. Radiation, pollution, and harsh weather can weaken skin over time. The quantum network senses these influences long before they cause visible harm. When ultraviolet light generates unstable energy in the upper layers, the nano coherence patterns shift to neutralize its effects, reducing cellular stress. When cold air threatens to dry the surface, the quantum elements help regulate moisture and strengthen the protective barrier. The skin responds like a living adaptive suit, adjusting in real time to maintain comfort and safety.

This technology also improves communication between the skin and the nervous system. The skin contains numerous sensory receptors that detect touch, temperature, and pressure. The quantum layer enhances these signals by filtering out noise and increasing clarity. This means sensations become more accurate without becoming overwhelming. Individuals gain sharper awareness of their surroundings, improving reaction time and coordination while still maintaining comfort.

Healing becomes faster as well. Small injuries that once took days to heal now heal in minutes or hours. Larger wounds recover with less pain and reduced risk of infection. The skin's quantum intelligence directs immune cells only where they are needed, preventing excessive inflammation. This creates a cleaner healing process that supports long-term health.

A further advantage is protection against microbes. Many harmful organisms disrupt local energy patterns when they attach to the skin. Quantum coherence detects these disruptions instantly and activates defensive responses. This could include strengthening the surface layer, sending signals to activate the immune system, or releasing antimicrobial compounds stored within the nanostructure. Instead of waiting for an infection to develop, the skin prevents harmful organisms from gaining a foothold.

Over time, quantum skin maintains youthful structure and elasticity. It constantly repairs micro-damage that naturally occurs from movement, expression, and exposure. The result is skin that remains strong, smooth, and flexible throughout life. Aging slows because the skin does not accumulate the tiny scars and imperfections that normally build up over decades.

Smart quantum skin with nano-coherence patterns represents a future in which the body's surface becomes as intelligent and adaptive as its inner systems. It protects, heals, senses, and strengthens with quiet precision, creating a layer of defense that feels both natural and extraordinary. This kind of skin transforms daily living into a smoother, safer experience, where the body remains resilient and balanced no matter what the environment brings.

Skin that becomes a sensory map for temperature, pressure, and light spectra.

Skin that becomes a sensory map for temperature, pressure, and light spectra reveals a future in which the body interprets the world with new clarity. Human skin already contains receptors for heat, cold, touch, texture, and pain, but their range is limited. Quantum-enhanced sensory skin extends these senses, using nanoscale structures and energy-sensitive elements to turn external stimuli into detailed information. This lets the body achieve a new awareness, making skin a map that reads the environment in depth.

Temperature becomes more than a simple feeling of hot or cold. Quantum structures inside the skin can detect variations so small that they occur before the nervous system would normally sense them. The skin picks up micro-changes in airflow, radiation levels, and surface temperature long before they cause discomfort. This allows the body to respond smoothly by adjusting blood flow, sweating, or muscle tension. It also improves safety, giving early warnings before burns, frostbite, or overheating can occur. The enhanced sensitivity works in harmony with natural thermal responses, making daily life more comfortable and reducing the risk of injury.

Pressure detection also transforms. In natural skin, pressure sensors send signals based on tissue deformation. Quantum sensory maps go far beyond this by sensing shifts in energy fields produced when an object approaches or touches the surface. This creates a more detailed picture of texture, weight distribution, and movement. For example, the skin can sense subtle vibrations from distant footsteps or detect the slightest shift in grip when holding an object. Everyday tasks become smoother and more precise, and activities that require fine motor control benefit from heightened accuracy. The sensory map also improves balance by giving the body more accurate information about ground contact and joint pressure.

Light becomes another source of information. Traditional skin does not directly detect light, aside from limited ultraviolet sensitivity. Quantum-enhanced skin, however, contains nanoscale elements capable of reading different wavelengths of light. The skin interprets infrared warmth, visible colors, and ultraviolet changes as part of its sensory landscape. This allows the body to sense the presence of nearby heat sources, changes in sunlight intensity, or shifts in ambient color without relying solely on eyesight. While it does not replace vision, it adds a new layer of environmental perception that supports safety, coordination, and situational awareness.

The sensory map also benefits emotional and social communication. Humans already express feelings through subtle skin changes—warmth, tension, goosebumps. With quantum enhancement, the skin can interpret similar cues from the environment or other individuals. Tiny changes in air pressure from movement, or shifts in temperature caused by emotional states, become clearer. This enriches human interaction and deepens intuitive understanding between people.

Another important aspect is how this sensory map communicates with the nervous system. Instead of overwhelming the brain with constant signals, the quantum layer filters and organizes information. It highlights meaningful changes while ignoring background noise. This creates a steady flow of useful data without causing sensory overload. The result is a body that feels more connected to its surroundings while remaining calm and balanced.

These enhancements also support safety in extreme environments. Whether navigating in the dark, handling delicate materials, or working near heat sources, the body maintains a detailed sense of what is happening around it. The skin acts as an intelligent interface, guiding movement and alerting the individual to potential hazards.

Skin that serves as a sensory map brings together physics, biology, and advanced materials to elevate the human experience. It sharpens the senses without altering their natural character, allowing the body to move, react, and connect with greater confidence. This future creates a world where the outer layer of the body is not just a protective covering but an active, intelligent guide through every moment of life.

Photonic camouflage inspired by quantum optics.

Photonic camouflage inspired by quantum optics imagines a world where the human body or advanced suits can blend into the environment with remarkable fluidity. Unlike traditional camouflage, which relies on patterns or colors that mimic surroundings, photonic camouflage works by controlling how light interacts with the surface. It bends, redirects, or absorbs light in ways that make objects appear invisible or seamlessly merged with the background. This idea stems from the same principles that govern how photons interact with atoms at the smallest scales, enabling materials to manipulate light with precision far beyond anything used today.

In this system, the surface becomes an active layer rather than a static design. Tiny quantum-based components adjust their structure in real time to match the lighting conditions of their environment. When standing in a forest, the surface reflects greens and browns in the exact arrangement seen behind the individual. In a desert, it reflects sand-colored wavelengths. In darkness, it softens reflections, eliminating outlines. This creates the effect of melting into the surroundings, as if the surface is no longer a separate object but a continuation of the space itself.

One of the most fascinating aspects of photonic camouflage is its speed. Natural chameleons change color slowly through chemical reactions in their skin. Quantum-inspired systems work through light manipulation, meaning adjustments occur at the speed of photons. The surface can shift in appearance at the slightest change in the environment. If a shadow passes overhead or the person moves from sunlight to shade, the camouflage adapts instantly. This responsiveness provides complete fluidity, avoiding the lag that would otherwise reveal a hidden presence.

Quantum optics also allows for more complex forms of camouflage. Instead of simply copying background colors, the system can reconstruct entire light pathways around the object. This means light travels as if nothing were there, creating a true invisibility effect. Observers see the scenery behind the surface even when looking directly at it. Unlike mirrors or reflective materials, this process does not create distortions or shine. It produces a smooth continuation of the environment that is difficult for the eye or camera to detect.

Another advantage is direction-based control. Photonic camouflage can adjust its appearance based on the angle of observation. If someone views the object from one direction, the surface recalculates the background pattern based on that line of sight. This makes detection extremely challenging because there is no single static pattern; the appearance changes with every viewpoint. This technique mimics how quantum particles behave, responding differently depending on whether they are observed.

Beyond invisibility, photonic camouflage provides protection from extreme conditions. By controlling how light is absorbed or reflected, the surface can regulate temperature. It may reflect harmful ultraviolet rays during intense sunlight or trap warmth in freezing environments. The same principles that hide the wearer also keep them comfortable and shielded from environmental extremes.

This technology can also be used for communication. Light patterns can shift subtly to signal teammates without producing sound or revealing position. These signals are invisible to anyone who does not know how to interpret them, adding an extra layer of strategy and safety.

The concept of photonic camouflage paints a picture of surfaces that act more like living systems than passive materials. They sense light, react instantly, and reshape their appearance with quiet intelligence. Whether used for exploration, security, wildlife study, or military applications, the effect is transformative.

Photonic camouflage inspired by quantum optics brings forward a future where the boundary between the body and the environment softens. The world becomes a place where visibility is no longer fixed but fluid, shaped by the same laws that govern the smallest particles of light. This creates a level of adaptability and stealth that feels both futuristic and entirely natural.

Applications in medicine, cosmetics, combat, and prosthetics.

Applications in medicine, cosmetics, combat, and prosthetics take on a completely new dimension when photonic and quantum-based technologies become part of the human body. These advancements turn the skin and the body's outer layers into intelligent surfaces that respond to light, pressure, and environmental changes with remarkable precision. Each field benefits in unique ways, creating solutions that feel both futuristic and practical.

In medicine, photonic and quantum-enhanced surfaces open the door to safer and more effective treatments. Doctors could examine injuries or internal issues without cutting into the skin, because smart surfaces can shift transparency on command, revealing underlying tissues through controlled light manipulation. This would allow real-time diagnosis without discomfort or invasive procedures. Patients recovering from burns or wounds would benefit from quantum-driven repair layers that rapidly reorganize cells, preventing scarring and reducing healing time. These surfaces can also monitor temperature changes, bacterial growth, and moisture levels, giving continuous feedback to ensure proper recovery. They act as protective barriers that adapt to the patient's needs, maintaining a steady healing process and reducing the risk of complications.

In cosmetics, the possibilities grow even more creative. Skin that adjusts its tone, texture, and brightness allows individuals to naturally change their appearance. Instead of makeup that must be applied and removed, the skin itself becomes a living canvas. Quantum-responsive pigments shift in response to light, highlighting natural features or softening imperfections. Tiny photonic structures alter how the skin reflects color, creating smoother, more radiant appearances without any external product. People could choose a brighter complexion for an event, a more subtle tone for daily life, or completely unique patterns for artistic expression. This offers long-lasting beauty solutions that adapt without harming the skin's health.

Combat applications leverage the technology's defensive and adaptive nature. Photonic camouflage can make soldiers nearly invisible by blending into their surroundings with unmatched accuracy. Instead of heavy gear, they would use lightweight suits that instantly adjust to terrain and lighting. These suits could also redirect heat signatures, making them hard to detect by thermal imaging. In dangerous environments, the smart skin can thicken itself upon impact, absorbing shock from explosions or protecting against sharp objects. Enhanced sensory feedback allows soldiers to detect vibrations, shifts in air pressure, and approaching movement through the skin, providing early warnings during missions. The technology becomes a protective partner, increasing safety while reducing physical strain.

Prosthetics also benefit significantly from these advances. Artificial limbs covered in photonic skin would look and behave like natural flesh. They would change color, match body temperature, and respond to touch with realistic sensitivity. Instead of a rigid surface, the prosthetic takes on the warmth, softness, and detailed appearance of living skin. This greatly improves comfort and emotional confidence for the wearer. Quantum sensors embedded in the prosthetic surface detect pressure, temperature, and movement, enabling the user to have intuitive control. The limb feels more connected to the body because the skin reads and translates environmental cues with precision. This makes every motion smoother and more natural.

Across all these fields, the core strength of quantum and photonic skin lies in its adaptability. It supports healing, enhances beauty, protects in dangerous situations, and restores a sense of wholeness to those using prosthetics. This technology brings together physics, biology, and design to improve both function and experience. It shows how the body's outer surface can become one of its most intelligent features, adjusting to life's demands with grace and precision.

CHAPTER 11

QUANTUM ORGANS – REPLACING THE BIOLOGICAL MACHINE

Quantum-powered artificial hearts, livers, and lungs represent a future where the most vital organs of the human body no longer struggle under stress, age, or disease. Instead, they operate with precision, stability, and resilience that go far beyond the limits of natural biology. These organs are not mechanical machines as imagined in the past, nor are they simple replicas of human tissue. They are hybrid systems shaped by advanced materials, guided by quantum sensing, and supported by microscopic structures that adjust themselves moment by moment. Together, they create a level of life support that feels natural yet performs with unmatched strength.

A quantum-powered artificial heart works by reading the body's energy rhythms rather than depending solely on mechanical pumps. The natural heart beats by responding to electrical signals that guide the timing of each contraction. These signals sometimes falter due to stress, illness, or aging. A quantum heart, however, senses these signals at a much deeper level. It detects the smallest shifts in the electromagnetic fields surrounding cardiac cells. Even before the body shows signs of irregular heartbeat or fatigue, the artificial organ adjusts its rhythm to maintain a steady flow of blood.

This precision prevents the heart from overworking or slipping into dangerous patterns. Because the organ responds instantly to changes in the body—such as exercise, stress, or emotional shifts—it delivers the exact amount of pressure and circulation needed at any moment. The heart does not tire. It does not weaken over time. It maintains smooth, efficient pumping regardless of age or physical condition. This means a person could run, climb, or endure intense physical activity with far less strain on the cardiovascular system than a biological heart would allow.

Another powerful feature comes from how the quantum heart distributes energy. Instead of relying on broad electrical pulses, it uses quantum coherence to coordinate contractions. This leads to smoother blood flow, reduced arterial wear, and a far lower risk of clot formation. The organ monitors blood thickness, oxygen levels, and vascular resistance, adjusting at every beat to match the body's state. It essentially becomes a living regulator that never loses balance.

The artificial liver benefits even more from quantum technology. The natural liver performs hundreds of tasks—filtering toxins, balancing hormones, storing nutrients, and processing chemicals. These processes often slow down as the body ages, and they are easily disrupted by disease. A quantum-powered liver does not face these limitations. It identifies toxins by reading their energy signatures rather than waiting for chemical reactions to develop. This allows it to neutralize harmful substances long before they reach dangerous levels.

Quantum-enhanced filtration surfaces separate compounds with incredible accuracy. They can distinguish between beneficial molecules and harmful ones even when they differ only at the atomic level. This reduces the burden on the rest of the body. The artificial liver also manages metabolism with more precision. It monitors glucose, fatty acids, and amino acids in real time and adjusts their levels based on the body's needs. This stabilizes energy supply, prevents spikes that lead to fatigue, and supports overall immunity.

In addition, the quantum liver does not accumulate damage the way biological livers do. Alcohol, pollutants, viruses, and medications often leave scars on natural liver tissue, eventually causing long-term harm. Quantum livers, built from self-repairing nanomaterials, regenerate themselves continuously. Their internal systems realign microscopic structures using coherence patterns that maintain perfect function. As a result, the liver remains strong and clear even after years of use.

The artificial lungs complete this trio of advanced organs. Natural lungs rely on air pressure and tiny alveoli to transfer oxygen into the blood and remove carbon dioxide. This delicate process can be interrupted by pollution, infection, inflammation, or chronic disease. A quantum-powered lung avoids these weaknesses through enhanced surface design and energy-based filtration. Instead of relying solely on airflow, it uses photon-sensitive membranes to attract oxygen molecules and push them into the bloodstream with high efficiency.

These membranes read the energetic behavior of gas molecules and sort them instantly. Oxygen passes through with ease, while harmful gases and particles are blocked or neutralized. This leads to cleaner breathing, even in environments filled with smoke, dust, or chemical pollutants. The lungs adjust their performance according to activity. During rest, they maintain a calm exchange of gases. During exertion, they expand their filtration capacity and speed up oxygen delivery, keeping the body energized without overworking the respiratory system.

The artificial lungs also protect against airborne pathogens. Many viruses and bacteria reveal themselves through tiny electromagnetic signatures. Quantum membranes detect these signatures before the organism enters deeper tissues. The pathogens are trapped, broken down, or reported to internal immune systems for immediate action. This creates a respiratory defense far stronger than anything found in nature.

Across all three organs, the greatest advantage is their ability to communicate with one another. Quantum systems share information not through slow chemical messengers but through fast, coherent energy signals. If the lungs detect rising carbon dioxide levels, they alert the heart instantly. If the liver senses a toxin entering the bloodstream, it signals both the heart and lungs to adjust circulation and breathing. This synchronized intelligence creates a body that stays balanced under pressure.

Another remarkable benefit is durability. Traditional artificial organs wear down, clog, or malfunction over time because their mechanisms are rigid. Quantum organs are fluid in structure. Their nanoscale components rearrange themselves to continue functioning even under stress or damage. Instead of declining with age, they maintain stability year after year.

These innovations also offer people with chronic illnesses greater freedom. A failing heart, scarred liver, or damaged lungs often requires constant medication or intense medical supervision. With quantum-powered replacements, individuals regain independence. They can lead active lifestyles without being concerned about unexpected organ failure or long-term complications.

Quantum-powered artificial hearts, livers, and lungs are more than replacements. They represent a new stage in human evolution—organs that protect, adapt, and strengthen the body in ways nature alone could never achieve. They offer stability, resilience, and performance, opening new possibilities for longevity and health. In this future, the body becomes not just restored but reimagined, capable of moving through life with power, balance, and enduring vitality.

Organs built from quantum-active materials that generate their own power.

Organs built from quantum-active materials that generate their own power introduce a future where the human body becomes a self-sustaining system, no longer limited by biological fatigue or the slow decline that comes with age. Instead of relying entirely on the body's natural metabolism, these organs pull energy from their surroundings, convert it into usable power, and operate with steady efficiency. They are not mechanical devices that need external charging, nor are they replacements that simply mimic nature. They are living, adaptive systems built from materials that respond to the smallest movements of particles, turning quantum activity into a constant source of strength.

The idea begins with materials that behave differently from anything in the natural world. Quantum-active structures do not rely on chemical energy the way traditional organs do. Instead, they draw power from physical processes that happen everywhere—vibrations in atoms, faint electromagnetic fields, ambient heat, and even the natural motion of molecules within the body. These materials convert such subtle energy into electrical or biochemical power, creating an organ that supports itself without exhausting the body's resources. The heart, the liver, the kidneys, and even small glands could, in theory, run on their own internal energy supply.

This self-charging capability is enabled by quantum-active materials, which respond instantly to environmental changes. At the smallest scales, particles are never still. They oscillate, collide, and shift constantly. When an organ is made from materials that can harvest energy from these movements, it becomes capable of generating power with every microscopic motion. Even body heat becomes a fuel source. Temperature changes cause atomic vibrations, and the organ transforms that motion into usable energy. This means power is produced through normal living—breathing, moving, thinking, and simply existing.

The benefits of this technology become clear when imagining a heart built from such materials. A natural heart requires a constant supply of ATP from cells, and this demand rises during stress or exertion. A quantum-active heart does not draw heavily from the body's resources. It creates its own power, ensuring that it never tires. Whether climbing a mountain, running a marathon, or experiencing intense emotional moments, the heart maintains stable performance. This reduces strain on the rest of the body and lowers the risk of cardiovascular fatigue, weakness, or failure.

A liver built from these materials could operate with similar efficiency. The liver performs energy-intensive tasks, including detoxifying chemicals, breaking down nutrients, and managing hormones. A biological liver often becomes overwhelmed under heavy stress, leading to fatigue or long-term damage. A quantum-active liver, however, produces its own power for its filtering systems. It uses quantum vibrations to fuel detoxification, allowing it to clear toxins faster without exhausting surrounding tissues. It processes nutrients more smoothly and keeps metabolic balance stable even under challenging conditions.

Organs that generate their own power also bring new possibilities for long-term health. Many illnesses arise because organs become weak or unable to keep up with the body's demands. A self-powered organ does not deteriorate in the same way. It does not depend heavily on the body's energy supply, so it does not cause strain on other organs. This means fewer cases of organ burnout, chronic fatigue, or metabolic imbalance. It also means the body can maintain performance even in environments with low energy intake, such as during illness, recovery, or extreme physical exertion.

Another fascinating aspect is the organ's ability to share excess energy with surrounding systems. When a quantum-active organ generates more power than it needs, it can send the surplus to nearby tissues. This gives an additional layer of support to muscles, nerves, and immune cells. During moments of stress, the body can draw on this extra energy to improve reaction time, strengthen defenses, or speed up healing. Over time, this kind of internal support could reduce the impact of aging, strengthen resilience, and accelerate recovery from injury.

The structure of these organs is equally impressive. Quantum-active materials can rearrange themselves at the atomic level in response to internal signals. This allows them to automatically repair wear and tear. A natural organ gradually accumulates damage through years of use, but a quantum-organ repairs micro-tears and structural imbalances as soon as they appear. It maintains its form with incredible accuracy, preventing many of the age-related conditions that affect traditional organs. This self-repairing process occurs quietly and continuously, helping the body stay strong and functional for decades.

Another transformative feature lies in communication. While biological organs rely on chemical messengers and slow electrical signals, quantum-active organs use coherent energy signals that travel much faster. They communicate instantaneously, adjusting their functions in perfect sync. When the lungs detect a shift in oxygen levels, the quantum heart responds immediately. When the liver senses a chemical imbalance, the kidneys adjust filtration without delay. This harmony keeps the body balanced in ways classical biology cannot match.

There is also the incredible potential for improved sensory awareness. Quantum materials are sensitive to electromagnetic and vibrational changes. An organ built from them might pick up early signals of illness, stress, or environmental danger. This information can be sent to the body, allowing corrections long before problems escalate. The organs become active guardians, constantly scanning and responding to internal conditions.

In addition to biological applications, these self-powered organs open new freedoms for people with chronic illnesses. Patients who depend on dialysis, oxygen support, or transplants would benefit from organs that do not fail, do not fatigue, and do not depend on outside energy sources. Instead of living with constant medical supervision, they could reclaim independence. A quantum-powered organ works quietly and consistently, removing the fear of sudden failure.

This vision also influences human capability outside medicine. Adventurers, astronauts, and individuals in extreme environments could survive longer with organs that maintain performance even when food, oxygen, or rest is limited. The body becomes more adaptable, more resilient, and better able to endure hardship.

Organs built from quantum-active materials that generate their own power represent a new chapter in human evolution.

They expand the body's capacity, protect it against decline, and maintain stability through intelligent energy use. These organs do more than replace damaged tissue—they redefine what it means to be strong, balanced, and durable. The body becomes a self-sustaining system, supported by materials that work elegantly at the smallest scales and guided by a form of intelligence shaped by physics itself.

Bio-quantum synchronization to integrate artificial organs with living tissue.

Bio-quantum synchronization offers a future where artificial organs and living tissue blend so seamlessly that the body cannot distinguish between what it was born with and what was added later. Instead of treating artificial organs as foreign objects, this approach helps them function as natural extensions of the human system. The key lies in aligning the energetic patterns of both biological cells and quantum-built components so they interact with perfect harmony. When this synchronization occurs, every part of the body communicates smoothly, heals together, and adapts as a unified structure.

At the core of this idea is the recognition that living tissue operates through constant energy communication. Cells send information not only through chemicals or electrical impulses, but through faint electromagnetic rhythms and vibrational patterns. These signals help tissues coordinate growth, signaling, immunity, and repair. Traditional artificial organs do not match these rhythms. They work mechanically or electronically, which creates a gap between the synthetic and the natural. Even the best modern implants struggle with long-term acceptance because the human body senses that their signals are different.

Bio-quantum synchronization solves this problem by enabling artificial organs to tune themselves to the body's energy field. Quantum materials inside the organ can read the frequencies of surrounding cells and adjust their own patterns accordingly. Just as a musical instrument can tune itself to match the pitch of another, the artificial organ aligns with the body's natural rhythm. This synchronization creates a stable relationship where the tissues merge with confidence rather than resistance.

The connection begins the moment the artificial organ enters the body. Quantum sensors built into the structure scan the surrounding cells, identifying their electromagnetic signatures. Each person has a slightly different energetic pattern, influenced by metabolism, age, emotional state, and health. The artificial organ adapts to this pattern, using quantum coherence to mimic the same rhythm. As a result, nearby cells recognize the organ not as an invader but as a partner. This reduces inflammation, lowers the risk of rejection, and promotes immediate healing at the integration site.

Over time, the bond deepens. As the body changes—through stress, rest, exercise, illness, or growth—the artificial organ adapts. Instead of remaining rigid, it shifts its energetic expression to match the new state. This flexibility is something modern implants cannot achieve because they are built from static materials. Quantum-active structures, however, behave more like living tissue. Their energy flows continuously adjust, allowing them to remain synchronized with the body throughout a person's life.

This type of harmony also improves communication between the organ and the nervous system. Nerves do not need forceful electrical stimulation to interact with the artificial organ. Instead, they use subtle signals that quantum materials can interpret instantly. When the brain sends instructions, the artificial organ responds with perfect timing. When the body's internal sensors detect changes in temperature, blood chemistry, or pressure, the organ reacts with equal accuracy. This creates a natural experience for the individual—movements feel fluid, breathing feels effortless, circulation feels smooth, and digestion feels balanced.

Another important result of bio-quantum synchronization is improved healing. When living tissue and quantum structures align, they exchange repair signals much more efficiently. If a small injury develops at the interface, the artificial organ detects it via energy distortion and provides immediate support. It might release healing compounds, adjust structural tension, or guide new cell growth to mend the area. This prevents complications and extends the lifetime of both the artificial organ and the surrounding biological tissue.

The synchronization also stabilizes the immune response. Normally, artificial implants trigger a low-level, ongoing conflict between the immune system and the foreign material. With quantum alignment, the immune system receives signals that match the body's internal pattern. It no longer treats the organ as a threat. Instead, immune cells interact with it peacefully, even assisting in long-term maintenance. This reduces chronic inflammation, which is a major cause of implant failure in current medical practice.

Bio-quantum synchronization also supports emotional and mental well-being. The human body responds strongly to energetic harmony. When tissues communicate smoothly, the nervous system remains calmer. The person feels more grounded, less fatigued, and more balanced. Artificial organs that align with the body help maintain this equilibrium. They do not disrupt internal rhythms, thereby contributing to a stable emotional state. This connection highlights the importance of viewing artificial organs not as mechanical parts but as living systems that participate in the body's overall harmony.

Another valuable benefit is adaptability. If the artificial organ needs upgrading or modification in the future, bio-quantum synchronization helps the new components integrate quickly. The organ already understands the person's energetic blueprint, making updates smoother and far less stressful. This is essential for long-term medical improvement, where people may need enhancements over decades instead of single implants for life.

This technology also supports complex multi-organ systems. When several artificial organs operate in the same body, quantum synchronization keeps them connected. They exchange energy information with each other just as natural organs do. If the artificial heart adjusts its rhythm, the artificial lungs modify their oxygen output. If the artificial liver detects toxins, it signals the heart and kidneys to increase circulation and filtration. This cooperation creates a more stable internal environment than any single organ could achieve on its own.

In individuals who require multiple replacements—heart, kidney, liver, or lungs—bio-quantum synchronization allows the entire system to operate as one unified structure. Instead of struggling with mismatched signals, the body enjoys smooth, continuous coordination. This reduces long-term complications and increases the quality of life.

Bio-quantum synchronization also shapes how artificial organs will influence the future of medical care. Surgeries become safer because the integration process is smoother. Recovery time is shorter because tissues accept the organ more quickly. Lifelong immune-suppressant medication becomes unnecessary. People with chronic illness regain confidence because their artificial organs support them fully without conflict or instability.

Most importantly, this technology transforms artificial organs from simple replacements into living partners. They cooperate with cells, listen to the body, respond to changes, protect internal balance, and maintain harmony throughout life. This creates a world where artificial organs do not feel artificial—they feel like an extension of the person's own biology.

Bio-quantum synchronization marks a turning point in medicine. It brings together the intelligence of living tissue and the precision of quantum physics, forming a bond that gives the body strength, resilience, and unity at every level.

Path toward full-body replacement and the ethical questions it raises.

The path toward full-body replacement begins with small steps—better prosthetics, stronger artificial organs, smarter sensory systems—but it eventually leads to a question that humanity has never faced before: what happens when nearly every part of the body can be rebuilt, improved, or replaced? The idea no longer belongs strictly to science fiction. As quantum materials, self-powered synthetic organs, and intelligent bionic systems evolve, the possibility of replacing most or even all biological structures becomes real. This future opens extraordinary opportunities for health, longevity, and human capability, but it also raises deep ethical questions about identity, equality, and the nature of human life.

The journey starts with necessity. People who suffer from organ failure, traumatic injury, or degenerative disease already depend on technology to survive. Artificial hearts, limbs, and sensory implants have transformed countless lives. Quantum-based systems push this even further, offering replacements that do not simply restore function but exceed it. A quantum heart that never tires, lungs that filter toxins effortlessly, or bones reinforced with graphene lattices turn weakness into strength. For many individuals, these advances mean freedom from suffering and the return of abilities they feared were lost forever.

As these technologies improve, they stop being simple replacements and become enhancements. A synthetic limb with quantum micro-motors delivers greater endurance than a natural arm. Skin with photonic camouflage can protect itself and respond to danger. Artificial organs that generate their own power reduce strain on the rest of the body. Eventually, the distinction between therapeutic technology and performance enhancement blurs. If a replacement can outperform the biological version, people may choose it not because they are sick, but because they want to achieve more.

This is where the path to full-body replacement becomes clear. A future human body could contain an artificial heart, lungs, liver, kidneys, bones, muscles, and sensory layers—all synchronized with living tissue through quantum communication. At that point, how much of the original biology remains? And does it matter? A body made from quantum-active materials would not age the same way. It would withstand extreme environments, resist disease, and constantly repair itself. Lifespan would stretch beyond natural limits, and physical limits would shift dramatically.

However, the closer society moves toward this possibility, the more urgent the ethical conversation becomes. The first question is identity. If the body is replaced piece by piece, is a person still the same individual? Most people already accept that a heart transplant or artificial limb does not change who someone is. But if nearly the entire body becomes artificial, the answer becomes less obvious. Does identity come from the mind, the continuity of experience, or the physical form itself? If consciousness remains intact, many would argue that the person remains the same. Yet others might feel uneasy at the thought of moving further from natural biology.

Another concern is inequality. Advanced synthetic organs and quantum-based replacements could be expensive at first. If only wealthy individuals have access to stronger bodies, longer lives, and enhanced abilities, society could become divided in a way that mirrors the deepest fears about technological progress. The gap between those who can afford the future and those who cannot may grow wider. Ethical discussions must address how access should be managed so that these advances benefit everyone rather than creating new forms of discrimination.

There is also the question of autonomy. If artificial organs rely on software or quantum programming, who controls that programming? Could an individual's body be influenced by external systems without their consent? Preventing misuse becomes as important as developing the technology itself. Clear rules would be necessary to ensure that people retain full control over their enhanced bodies.

Some thinkers question the meaning of human limits. Pain, fatigue, and aging have shaped human culture, values, and psychology for thousands of years. If these limits fade, will people approach life differently? Will accomplishments still feel meaningful when the body can endure almost anything? A world of near-indestructible bodies may offer freedom, but it also changes the relationship between effort, struggle, and achievement.

Environmental impact also enters the discussion. Artificial bodies require materials, energy, and maintenance. What happens when millions of people choose full or partial replacements? The technology must support long-term sustainability to avoid creating new ecological burdens.

Despite all these concerns, many see full-body replacement as an extension of something deeply human: the desire to heal, adapt, and improve. Humans have been modifying the body for centuries through medicine, prosthetics, exercise, and lifestyle. Technological evolution simply expands what is possible. Instead of accepting disease or decline as inevitable, future generations may choose bodies that remain strong and capable throughout life.

The ethical questions are not meant to stop progress but to guide it wisely. As artificial organs blend with living tissue through bio-quantum synchronization, and as the line between replacement and enhancement fades, society must decide how to preserve dignity, fairness, and autonomy. Discussions involving scientists, ethicists, lawmakers, and the public will shape the future path.

Ultimately, the idea of full-body replacement challenges humanity to examine what truly defines a person. Is it the physical form, the memories carried in the mind, the relationships shared with others, or something deeper that cannot be replaced by any material? The answer may vary from one individual to another.

The path ahead promises extraordinary opportunities. It offers relief from suffering, freedom from disease, and bodies capable of extraordinary performance. Yet it also demands thoughtful choices about how to protect what makes life meaningful. Full-body replacement is not just a technological milestone; it is a turning point where humanity must decide how to evolve while honoring the essence of being human.

CHAPTER 12

QUANTUM ENERGY – POWERING THE BIONIC HUMAN

Harvesting energy from quantum fluctuations and biophoton emissions paints a picture of a future in which the human body no longer depends solely on food, oxygen, or chemical reactions for strength. Instead, it taps into subtle sources of power that are always present but currently unused. These sources exist at the smallest scales of reality, in the faint movements of particles, the natural glow emitted by cells, and the constant dance of energy within living tissues. By learning to capture and convert these hidden forms of energy, the body gains access to a renewable, steady, and deeply embedded power supply.

Quantum fluctuations are among the most fascinating of these sources. At the microscopic level, particles never stand still. Even in what appears to be total emptiness, they flicker in and out of existence, creating tiny bursts of energy. These fluctuations are incredibly small, but they occur everywhere and at all times. Technologies built from quantum-active materials could capture tiny amounts of this energy and convert it into useful power for artificial organs, bionic systems, and even everyday cellular functions. This creates a constant background charge that supports life without drawing heavily on biological resources.

Imagine a heart that uses quantum fluctuations to maintain part of its pumping force. Even if the body is tired, dehydrated, or under stress, the heart remains strong because it draws power from the underlying fabric of reality. Muscles could gain similar support, receiving small boosts during movement so they do not strain as easily. The body becomes more resistant to fatigue because it no longer relies only on chemical reactions to produce energy. Quantum fluctuations fill in the gaps, offering continuous assistance.

Bio-photon emissions provide another natural energy source. All living cells release tiny bursts of light as part of their normal functions. These bio-photons are extremely weak, invisible to the human eye, yet they carry important information about the body's state. Some scientists believe these emissions help coordinate cell communication. Technologies designed to work with this natural light could harvest it, amplify it, and convert it into usable energy.

A quantum-powered organ, for example, could read the bio-photon signals around it and capture the energy released during cellular activity. This does not weaken the cell. Instead, the organ uses the light signals—already present and already flowing—to support its own systems. In this way, the organ becomes a partner in energy production rather than a passive recipient of fuel. The more active the surrounding tissue, the more energy becomes available.

This creates a cycle of cooperation. Cells release biophoton energy, which is gathered by quantum structures. These structures convert energy into power to support organ functions or to send beneficial energy back into nearby tissues. The overall result is a body that smoothly transfers energy between systems, reducing waste and improving stability.

By capturing energy from both quantum fluctuations and bio-photon emissions, artificial organs and bionic systems reduce strain on the biological body. For example, the liver often works hardest when processing toxins or digesting large amounts of nutrients. A quantum liver that gathers ambient energy can perform at high efficiency without borrowing too much power from the rest of the body. The lungs process oxygen more smoothly when their membranes have a steady flow of energy from their quantum layers. Even small organs, such as glands, remain stable under stress because they have an alternative nutrient supply.

This form of energy harvesting also allows for better cellular repair. The body naturally repairs minor damage every day, but this process slows when energy levels are low. Quantum systems gather ambient energy and direct it toward areas that need healing. Micro-tears in muscle fibers close faster, and tissues recover more easily from strain. Even the immune system works more efficiently when supported by consistent energy flow from quantum harvesting.

The effects extend beyond healing. People who often feel fatigued due to stress, low nutrition, or irregular sleep schedules might experience greater stability. Their bodies would not crash as easily because the internal energy network would constantly replenish itself. Daily routines become smoother, mental clarity improves, and long-term health becomes easier to maintain.

This technology also opens exciting possibilities for survival in extreme environments. In space, deep underwater, or remote areas with limited oxygen or food, the body could rely on quantum fluctuations and bio-photon harvesting to maintain essential functions. Instead of being vulnerable to environmental extremes, the body becomes more self-sufficient. Explorers, astronauts, and individuals living in harsh climates gain a new level of independence and resilience.

There are also environmental benefits. A body powered partly by quantum sources reduces the need for external resources. It produces less metabolic waste because it does not rely exclusively on traditional fuel pathways. Over time, this could ease pressure on food systems and lower the physical toll that modern lifestyles place on the body.

Another remarkable aspect of this energy system is stability. Quantum fluctuations occur everywhere and never run out. Bio-photon emissions continue as long as cells remain alive. Together, they create a near-constant energy grid within the body. Artificial organs made from quantum-active materials would never completely power down unless intentionally designed to do so. They remain ready, responsive, and strong even under sudden shocks or shortages.

Ethical questions arise, of course. If the body becomes partially self-powered, what does that mean for metabolism, diet, or the natural limits of human performance? Will some people gain advantages that others do not have? How should society regulate enhancements that fundamentally change how humans draw energy from the world?

These questions are important, but they highlight how transformative this technology could be. Harvesting energy from quantum fluctuations and bio-photon emissions is not simply an engineering achievement—it is a shift in how the body interacts with reality itself. Instead of seeing energy as something consumed and depleted, the body becomes part of a steady, ever-present flow.

The result is a system that supports strength, endurance, healing, and stability in ways classical biology cannot match. It offers a glimpse of a future where humans are not just surviving but thriving through harmony with the deepest structures of the universe.

Zero-point energy concepts for powering implants without batteries.

Zero-point energy is one of the most intriguing ideas in modern physics, and it offers the possibility of completely reshaping how implants and artificial organs receive power. At the smallest levels of reality, even in a state that appears perfectly still, energy never truly reaches zero. Particles continue to move, vibrate, and fluctuate at a level so subtle that it seems almost imaginary. This constant activity is known as zero-point energy, and it exists everywhere—within the air, within materials, within the human body, and even within what appears to be empty space. If this hidden reservoir of energy could be tapped safely and consistently, implants would no longer need batteries, external chargers, or chemical fuel. They would operate through a power source that never runs out.

To understand how this becomes possible, picture the idea that space is not truly empty. Instead, it behaves like a restless stage where particles continuously appear and disappear. Even at extremely low temperatures, where classical physics predicts complete stillness, these particles still move. This movement produces tiny units of energy that fill every point in the universe. Engineers and scientists imagine constructing implants made of quantum-sensitive materials that capture small amounts of this background energy. The materials could be designed to resonate with the natural fluctuations around them, absorbing and converting them into useful electrical power.

An implant powered in this way would not need a battery that wears out over time. It would not rely on chemical reactions that fail under stress, nor would it require large internal components that take up space in the body. Instead, the implant works like a quiet receiver, absorbing the faint but constant pulse of zero-point energy and using it to run its internal systems. This gives the implant a kind of steady heartbeat that never stops as long as it remains within the fabric of space itself.

For many people with medical devices, power management is a major concern. Traditional pacemakers, for example, require occasional replacement when their batteries weaken. Neural implants often rely on external chargers, which can create discomfort or inconvenience. Artificial organs sometimes face limits because they depend heavily on the body's energy supply. Zero-point energy changes all of this by providing a power source that is constantly available, unavoidable, and not influenced by diet, oxygen levels, or physical activity. This provides implants with reliability that traditional systems cannot match.

A quantum-powered implant would also operate more smoothly. Because the energy supply is steady and continuous, the device avoids sudden drops in efficiency or unexpected slowdowns. For instance, a hearing implant that uses zero-point energy would remain fully functional even during intense exercise, stress, or illness. Its performance would not change whether the person is awake, asleep, or in an environment with limited oxygen. The implant simply draws energy from its surroundings to maintain its functions.

The safety factor is very important in making this technology appealing. Zero-point energy does not require radioactive materials or dangerous fuel cells. It also does not introduce heat that could damage tissues. Because energy is gathered in extremely small amounts, the implant uses only what it needs, making the process gentle. No excessive buildup occurs, and there is no risk of overheating. Instead, the power flow is precisely controlled, ensuring that tissues remain unharmed and that the implant functions at a natural rhythm.

One of the most exciting parts of this concept is how it interacts with other biological systems. When implants no longer depend on traditional power sources, they can become more compact. Smaller designs integrate more easily with living tissue and place less strain on it. A neural implant could be thin enough to sit within delicate areas of the brain without interfering with natural processes. A heart-support device could fit seamlessly into the chest without additional wires or bulky components. This creates a smoother relationship between technology and the biological body.

Another advantage is longevity. Zero-point-powered implants could last for decades with minimal wear because they do not contain moving parts that degrade or chemicals that lose potency. This significantly reduces the need for repeated surgeries or replacements. A single implant could support a person for most of their life, maintaining strong performance through all stages of aging. The reliability of such a device brings peace of mind to individuals with chronic conditions or severe injuries.

This technology also opens the door to advanced designs that go far beyond what traditional batteries can support. For example, implants that require rapid processing, high-energy bursts, or continuous data monitoring often face power limitations. With zero-point energy, these limitations disappear. A quantum prosthetic limb could operate with lightning-fast response times, reading signals from the nervous system with extreme accuracy and reacting almost instantly. Muscles in the artificial limb could function smoothly at all times because they receive constant power.

The same principle applies to artificial organs that perform chemical filtration or mechanical pumping. A quantum-powered liver implant could maintain high efficiency even under heavy biological strain. A lung implant could process air more consistently, providing stable oxygen flow regardless of environmental conditions. These solutions no longer depend solely on the person's metabolic state.

Instead, they become partly independent, forming a hybrid system that blends biological power with the subtle strength of quantum fluctuations.

Beyond personal health, the concept has far-reaching implications for environments with scarce energy resources. In deep-space missions, where access to sunlight or electrical power can be limited, implants that rely on zero-point energy create a safer, more predictable environment for astronauts. Devices remain active and reliable even when external conditions change abruptly. This stability becomes essential for long journeys where maintenance opportunities are rare.

Although the idea is powerful, it also raises scientific and ethical questions. How much energy should an implant be allowed to harvest at once? Could large-scale harvesting affect the environment, or would the impact be too small to notice? How should the technology be regulated to prevent misuse? These questions highlight the need for careful development and responsible use. While the energy at the quantum level is immense, the harvesting process must remain controlled and safe.

The potential for misuse also exists. If implants become too advanced, inequalities may form between those who can access the technology and those who cannot. Therefore, systems must be designed to support fairness and safety while avoiding unintended consequences.

Despite these concerns, zero-point energy remains one of the most promising ideas for powering future medical devices. The constant presence of this energy and the ability to convert it with quantum-sensitive materials create a new foundation for healthcare. Instead of relying on batteries that run out or technologies that require constant monitoring, implants become self-sustaining partners in the human body.

In the long term, this approach could transform the meaning of medical support. Devices would not be temporary aids but long-lasting components that function as reliably as the body's own organs. The union of physics and biology would allow humans to carry powerful tools within their bodies without ever worrying about charging or replacement. Through the gentle collection of zero-point energy, implants would gain a quiet, dependable power source that continues for as long as space itself remains active.

Quantum-efficient metabolism and enhanced energy conversion.

Quantum-efficient metabolism is a vision of the human body processing energy with a level of precision and speed far surpassing the limits of ordinary biology. It suggests a system in which every calorie is converted with almost no waste, every molecule is used to its fullest potential, and every cellular reaction operates at the highest level of efficiency. This idea becomes possible when biological processes work together with quantum principles, allowing the body to handle energy not only through chemistry but also through the subtle movements of particles, vibrations, and energy states at the smallest scales of nature.

At the heart of this concept is the idea that the human body loses substantial energy in the form of heat, friction, and incomplete chemical reactions. Traditional metabolism, even at its best, wastes significant resources. The body uses food to create energy, but much of that potential remains unused because chemical reactions do not always align perfectly. Enzymes sometimes react more slowly than needed. Cells burn fuel inefficiently. Electrical signals face resistance. All these minor losses add up over time, limiting strength, endurance, and overall health.

Quantum-efficient metabolism works differently. Instead of relying solely on large chemical reactions, it leverages quantum behavior within molecules. In this model, electrons move through pathways that allow near-zero resistance, and energy transfers occur with minimal loss. Reactions inside cells align more precisely because electrons and molecular vibrations move in coordinated patterns. This coordination means the body can extract more usable energy from the same amount of food, while producing fewer byproducts that slow down or stress the system.

The key idea is coherence. When electrons in biological structures move in a synchronized state, energy flows far more smoothly. This is similar to how quantum particles in certain materials can move collectively without resistance. When cells operate in this synchronized rhythm, reactions speed up, energy waste drops, and the entire metabolic chain becomes stronger and more balanced. Even small improvements in efficiency here could lead to significant changes in physical performance and overall well-being.

Imagine muscles that receive energy almost instantly, thanks to fast, efficient quantum tunneling. Instead of waiting for slow chemical steps, the body accesses stored energy directly through synchronized electron movement. This gives cells quicker access to fuel during activities such as running, lifting, or recovering from injury. Muscles feel less fatigue because they waste less energy and regenerate ATP more rapidly.

Another part of this idea involves mitochondria, the structures that convert food into usable energy. With quantum-enhanced behavior, mitochondria could operate with much higher precision. They could detect energy demands earlier, adjust their output more quickly, and reduce the production of harmful byproducts, such as free radicals. This would lower inflammation, slow down cellular aging, and support better long-term health. Many age-related issues stem from slowed or impaired energy production, and quantum-efficient metabolism offers a way to reduce this decline.

This approach does not stop at energy production. It also improves how the body stores and distributes energy. Cells could maintain tiny energy reserves in quantum-stable states, ready to release when needed. These reserves could respond to stress or sudden movement without delay. This means quicker reflexes, smoother coordination, and faster adaptation during physical challenges. The body would feel more responsive, stable, and balanced because energy is always available at the exact moment required.

A quantum-efficient system might also make digestion smoother. Instead of relying on slow chemical breakdown, the body could guide molecules into more efficient reaction paths. Enzymes could process nutrients with higher accuracy, selecting the most useful components and directing them precisely where they are needed. This reduces the strain on organs such as the liver and pancreas, which often work overtime to handle large or complex meals. As a result, the body could feel lighter, cleaner, and more energized after eating, even when the meal is heavy.

Another interesting part of this idea is heat management. In normal biology, heat is one of the biggest forms of wasted energy. Muscles generate heat during exercise, and the body must use energy again to cool itself. With quantum-efficient metabolism, this heat production decreases because reactions produce less waste. Internal temperature stays more stable, reducing fatigue and improving endurance. Athletes and individuals under physical stress could perform longer without overheating.

The immune system also gains advantages. Cells that require energy to fight infections, repair tissues, or maintain protective barriers would receive a constant, steady supply. This supports a stronger, faster immune response, giving the body greater resilience. Energy-rich cells can act quickly, reducing the time it takes to eliminate threats or recover from illness.

Quantum-efficient metabolism could also influence the brain. Neurons rely heavily on energy to maintain electrical signals, repair themselves, and communicate. When metabolism becomes more efficient, the brain receives consistent support, improving clarity, memory, and focus. Faster energy availability helps neurons fire with precision, strengthening cognitive abilities and reducing mental fatigue. This could lead to sharper thinking, better emotional regulation, and an overall sense of stability.

In daily life, this kind of metabolism would make ordinary tasks feel lighter. Movement would require less effort. Sleep would feel more restorative because cells repair themselves more efficiently at night. Stress would cause less exhaustion because energy stores replenish quickly. The body would maintain a steady rhythm, adapting smoothly from rest to activity and back again.

The long-term impact of such efficiency reaches even deeper. With less wasted energy and fewer harmful byproducts, the aging process slows down. Cells maintain their structure longer. Organs function with greater balance. The risk of metabolic diseases decreases because the body better handles sugar, fats, and proteins. Even small changes in metabolic stability can influence longevity, and quantum-efficient metabolism presents a pathway for major improvements.

To achieve something like this, future bionic designs would need to work closely with natural cells, guiding them toward quantum-coherent behavior. Structures made from quantum-active materials could sit beside mitochondria, helping them tune their energy output. Nanodevices could regulate electron flow, ensuring it follows the most efficient path. These systems would act as silent partners to biology, enhancing natural responses without forcing the body to depend entirely on external machines.

While such advancements remain theoretical today, the potential is enormous. They offer a future where humans can rely on stronger internal systems, better health, and greater endurance. Quantum-efficient metabolism brings together the subtle rules of physics with the complex functions of cells, creating a new blueprint for how the body can process energy. It pushes the idea of human health beyond traditional limits, giving the body the power to operate with the precision of a finely tuned engine and the resilience of an advanced natural system.

Future vision of humans who never run out of biological "power."

A future where humans never run out of biological "power" paints a picture of life operating with smooth, steady energy at every moment. It imagines bodies that do not crash from exhaustion, minds that do not fade when tasks grow demanding, and systems that maintain strength without relying on constant food intake, stimulants, or rest. This vision builds on the idea that the body could use quantum principles to manage energy with unmatched precision, drawing on internal reserves and the surrounding environment in ways that traditional biology cannot. In this world, energy becomes less of a fragile resource and more of a continuous flow that supports every part of life.

Human energy today is shaped by limits. Muscles tire after exertion. The brain slows down after long periods of concentration. Organs use large amounts of fuel and lose efficiency over time. These limitations shape daily life, influencing mood, performance, and overall health. People struggle with fatigue, hunger, and mental strain because biology was built for survival, not for endless efficiency. But the introduction of quantum-enhanced metabolism and energy systems challenges these boundaries, creating a model in which the body can maintain peak performance for far longer.

In this imagined future, cells use energy with such precision that almost nothing is wasted. Instead of losing power to heat or inefficient chemical reactions, cells convert fuel into usable energy with near-perfect efficiency. Even small amounts of food can provide long-lasting energy because the body captures every bit of energy hidden in molecules. Mitochondria operate like finely tuned engines, adjusting instantly to meet physical or mental demands without delay. This level of energy control turns ordinary biological processes into powerful engines of endurance.

A person living in this future would experience a body that feels constantly ready. Morning fatigue, mid-day crashes, and evening slumps would no longer shape routines. Instead, the body would maintain steady energy from sunrise to night, adjusting seamlessly to stress, movement, and rest. Someone could read for hours without losing focus, walk long distances without tiring, or handle demanding work without feeling drained. The natural rhythm of life becomes smoother as the body supports every action with a stable supply of power.

This transformation would also influence emotional well-being. Many emotional difficulties arise indirectly from physical depletion. When energy levels fall, irritability grows, motivation decreases, and stress feels heavier. In a body with constant power, emotional balance becomes easier to maintain. The mind thinks more clearly, handles challenges with confidence, and remains grounded even during stressful moments. This steadiness strengthens decision-making and promotes a sense of inner calm.

Such a body would also recover rapidly from physical stress. Muscles repair themselves quickly because they receive abundant energy and resources. Damage from strain or injury heals faster as cells work more efficiently. This reduces the long-term effects of aging, as tissues degrade more slowly and maintain their strength. People could continue to move with agility and confidence well into old age, maintaining physical capabilities that typically fade with age.

In this future, the body depends less on food for survival. Hunger still exists as a natural signal, but the urgency behind it weakens because the body can stretch available nutrients across longer periods. Quantum-level processes allow cells to channel energy from molecular vibrations, ambient electromagnetic fields, or small environmental inputs. This does not replace food entirely, but it supplements it to prevent sudden drops in energy. People skip meals without losing strength, travel long distances without carrying large supplies, and endure emergencies with better resilience.

The brain becomes one of the greatest beneficiaries of this constant power. Every thought, memory, and movement relies on electrical impulses. When energy is abundant and steady, the brain operates with clarity and speed. Memory improves because neurons communicate more efficiently. Creativity expands as the brain has more resources for complex thinking. Attention remains strong even during challenging tasks. This leads to heightened mental performance, supporting every aspect of personal and professional life.

This future also allows for stronger integration between the body and advanced technology. Bionic organs, quantum implants, and enhanced prosthetics function far more effectively when they have unlimited access to power. These devices can operate without batteries, relying entirely on energy supplied through quantum-efficient biological systems. Artificial limbs move smoothly and precisely. Sensory implants provide vivid perception. Internal devices continuously monitor health without draining the body's energy reserves. The relationship between technology and biology becomes more fluid, creating a hybrid system where each supports the other.

Another fascinating part of this vision is the impact on extreme environments. People would handle harsh climates, long periods of physical activity, or high-pressure conditions with greater ease. Climbing mountains, exploring underwater regions, or traveling through remote landscapes would feel less risky because the body maintains strength even in tough situations. Astronauts exploring other planets could endure long missions without fatigue or energy loss. This steady supply of power expands what humans can accomplish beyond Earth as well.

The benefits of this future extend into personal lifestyle. People could pursue hobbies, sports, and creative activities without feeling constrained by physical limitations. Energy becomes a constant companion instead of a fluctuating obstacle. Exercise becomes more enjoyable because the body wears down less quickly. Long work shifts feel manageable. Parenting becomes easier because energy remains available even on stressful days. This creates a more balanced and fulfilling life experience.

However, such a future raises important questions. If humans never run out of power, how will society adapt? Would people take on more responsibilities than they can handle? Would constant energy lead to unrealistic expectations in workplaces? These questions remind us that energy is not only a physical resource but also a factor that shapes social structures, habits, and lifestyles. A future with limitless power requires careful consideration of how to maintain balance and well-being.

Despite these challenges, the core idea remains incredibly inspiring. A body that maintains constant biological power represents a step toward freedom from many physical struggles that shape life today. It reflects a future in which humans can perform with strength, clarity, and stability, guided by systems that harmonize with the deepest principles of nature. Quantum-efficient energy processing transforms what it means to be alive, offering a life filled with vitality, resilience, and potential.

CHAPTER 13

QUANTUM INTERFACES – THE HUMAN-MACHINE CONNECTION

Quantum entanglement-based communication enables instant, physics-rooted information exchange between devices and humans—regardless of distance—without the need for traditional signals. Here, communication happens through deeply connected particles: a change in one instantly affects the other. This creates a far faster, more secure interaction pathway. When used in bionic systems and advanced implants, it transforms how bodies and technology work together, enabling seamless, immediate information flow.

To picture this future, think about how a human brain and a bionic device interact. Current implants use electrical signals, tiny electrodes, or wireless systems to send and receive instructions. These work but suffer from delays, interference, and a limited range. Entanglement overcomes this by directly linking two systems. When particles are entangled, they act as one unit, no matter the distance. They mirror each other's states in real time. This allows communication without classical pathways.

If a device and the human body share entangled particles, the device can instantly interpret signals, such as thoughts or physical needs, without delay or translation. It connects to the user's internal state at the quantum level, enabling precise and immediate response.

For example, consider a quantum-controlled heart implant. Instead of waiting for nerve signals, the implant changes rhythm the moment the body's state shifts. It senses stress, relaxation, movement, or stillness through entangled particles. The heart beats in time because it gets information instantly, without delay. This creates harmony between the device and the body, enabling smooth coordination with every action.

In another scenario, a neural implant designed to enhance memory could store information and retrieve it instantly through entangled signals. When the brain begins to form a memory, the implant receives the information simultaneously. It organizes, strengthens, and keeps memory accessible without disrupting the brain's natural processes. This creates a cooperative relationship where human thought and artificial intelligence combine into a single, fluid system.

This method extends beyond internal systems, transforming how people and devices interact. Entangled connections allow control through intention; devices respond instantly and intuitively, enabling natural operation without traditional commands.

Security is another powerful aspect of entanglement-based communication. Because entangled signals do not travel through ordinary channels, they cannot be intercepted in the usual ways. Hackers cannot tap into them, and no one can record or duplicate the signals without breaking the entangled state. This creates a communication system that is nearly impossible to breach. In medical settings, this ensures that implants remain safe from outside interference. In technological environments, it provides a level of privacy that far surpasses conventional encryption.

Entanglement also allows multiple devices in the body to work together as a unified network. Today, different implants may operate separately because they rely on independent sensors or processors. Quantum communication allows them to share information instantly. A lung implant could coordinate with a heart implant to adjust breathing and blood flow. A neural device could signal a sensory enhancement implant about emotional stress, allowing the sensory system to appropriately calm or sharpen perception. This unity creates a more coherent experience where the body and its enhancements move as a synchronized system.

Another layer of this vision involves emotional intelligence. Human feelings create subtle changes in electrical patterns, chemical signals, and electromagnetic fields. Quantum-entangled implants could sense these shifts with remarkable accuracy. A device could understand when the user feels anxious, focused, relaxed, or overwhelmed and adjust its behavior accordingly. A memory support implant might soften stressful memories during high emotional strain. A sensory device might lower its intensity when emotions run high. These responses create a more compassionate and adaptive partnership between technology and the human experience.

Entanglement-based communication also expands possibilities for remote medical care. Doctors could receive real-time information from implants without delay, even over long distances. Lives could be saved during medical emergencies because data reaches physicians instantly. Surgeons could perform precise remote procedures using entangled feedback, allowing them to sense what the patient's body experiences as if they were physically present. This level of connection brings new meaning to the idea of global healthcare.

For individuals with disabilities, this communication system redefines independence. A person with limited mobility could control household devices, transportation systems, or communication tools through entangled signals. Their intentions could guide robotic assistants or exoskeletons that move with natural grace. The connection feels seamless because the device responds as quickly as the person thinks. This gives individuals greater freedom and empowers them to navigate life with confidence.

The future also holds the possibility of shared entanglement networks among people. Families, teams, or caregivers could connect through controlled entanglement links that allow rapid emotional or physical sensing. This does not mean sharing private thoughts, but rather creating a subtle awareness of each other's states. For example, a parent could sense when a distant child is distressed. A medical team could monitor patients during recovery. First responders could coordinate their actions without relying on radio communication, improving safety and efficiency.

Despite the immense promise, this future brings ethical and philosophical questions. What boundaries must exist between human thought and device interpretation? How much control should technology have over human intention? Should shared entanglement networks be limited to medical or emergency uses? These questions guide the responsible development of such systems, ensuring that technology enhances life without overshadowing personal autonomy.

Quantum entanglement-based communication makes distinctions between human intention, biological function, and advanced technology nearly vanish. Devices respond instantly and naturally, enhancing our lives with seamless, intuitive, and powerful connections through communication at the speed of nature's fundamental laws.

Qubit-driven wearables that sync with physiology in real time.

Qubit-driven wearables are the next step in merging technology with the human body. Unlike simple gadgets, they act as intelligent companions, constantly interpreting signals and adapting instantly. Given that qubits can be found in multiple states simultaneously, these wearables process information incredibly quickly, making them responsive and deeply attuned to human physiology. This marks the rise of devices that feel alive and adaptive.

In today's world, wearables record steps, monitor heart rate, and track sleep patterns, but they still have noticeable gaps. They collect information after it happens and deliver feedback later. They rely on fixed sensors that cannot catch every detail, and they base their decisions on limited processing power. Qubit-driven systems change this by reading biological signals in real time and responding instantly. Instead of waiting for patterns to form, they recognize them the moment they appear. The device accurately detects shifts in heart rhythm, changes in breathing, and fluctuations in stress. It reacts before the wearer even notices the change.

This creates a seamless partnership between body and device. Imagine a wearable wrapped gently around the wrist. Instead of gathering data in chunks, it continuously interprets subtle changes in sweat chemistry, electrical signals, temperature, and micro-vibrations. It does this without lag because qubits enable the system to process countless variables simultaneously. A classical chip handles data step by step, but a qubit processor can analyze overlapping states, giving a picture of the body that is both deep and immediate.

Such a device could help someone stay balanced throughout the day. When stress begins to rise, the wearable senses the shift instantly. It might adjust its temperature to create a calming sensation or release a soothing vibration pattern that helps stabilize the nervous system. If blood sugar begins to drop, the device could notify the wearer gently or communicate with other bionic systems to restore balance. If the heart shows early signs of strain, the wearable responds before discomfort becomes visible.

These systems go beyond physical health. They can support emotional well-being by recognizing early signs of anxiety or mental fatigue. Changes in breathing rhythm, subtle muscle tension, and shifts in skin conductivity all provide clues to emotional states. Classical devices can detect some of this data, but they cannot integrate it fast enough or interpret it with nuance. Qubit-driven wearables handle the complexity of human emotion with greater sensitivity, offering assistance before stress becomes overwhelming.

An exciting aspect of this technology is its relationship to movement. When the body prepares to move, the brain produces tiny electrical patterns that signal the muscles. Wearables powered by qubits can read these signals instantly and adjust their support. For athletes, this means smoother control, better balance, and enhanced coordination. The device learns the person's movement style, adapts to their goals, and supports performance without feeling like external machinery. For people in rehabilitation, the wearable serves as a guide, providing precision and helping them regain strength through targeted adjustments.

The relationship between qubit-driven wearables and the environment also plays a major role. These devices can sense changes in temperature, air quality, and electromagnetic fields with extraordinary accuracy. They translate this information into actions that protect the wearer. If the air becomes polluted, the wearable might activate a small filtration system or notify the person before the environment becomes harmful. If ultraviolet levels rise, it might shift its surface pigments to shield the skin. These subtle adjustments create a safer experience in daily life.

Another powerful advantage is communication. These wearables can interact with other devices through quantum-secure channels, ensuring that personal health data stays protected. Such security prevents hacking or unauthorized access, giving the wearer confidence that their information is safe. The device could communicate with cars, homes, workplaces, or medical systems, creating environments that respond intelligently to human needs. A home might adjust its lighting or temperature the moment the wearable senses that the wearer is tired or stressed. A workplace might adjust brightness or screen settings to reduce eye strain. This creates a living ecosystem that listens and adapts.

These wearables also enhance creativity and learning. When the brain enters a focused state, the wearable can detect it and adjust its support to maintain concentration. When the mind becomes fatigued, it can provide gentle cues that help restore clarity. This makes activities like studying, designing, or problem-solving feel more natural, as the device becomes a helper that understands mental rhythms.

For individuals with health conditions, the technology becomes transformative. People with heart issues, asthma, diabetes, or chronic pain gain a constant companion that watches over them. Instead of reacting to emergencies after they occur, the wearable predicts them early and acts immediately. For example, if it senses the beginning of an asthma flare, it could activate a mild intervention or alert caregivers before breathing becomes difficult. If blood pressure begins to rise, it could adjust posture through subtle vibrations or communicate with implants that regulate cardiovascular function.

The appearance and feel of these devices also evolve. They no longer need to be rigid or noticeable. Qubit-driven wearables can be woven into clothing, wrapped as thin films over the skin, or integrated into jewelry that feels natural and expressive. The technology disappears into everyday life, blending with personal style rather than standing out as a machine. Some versions might even be transparent, flexible, or self-healing, repairing small tears through quantum-coordinated material responses.

Despite the amazing benefits, this future raises important questions. How much should wearables influence human behavior? Should devices predict emotional states, or should they only respond when asked? These questions guide how society integrates quantum technology into personal life. The goal is a partnership, not control. Qubit-driven wearables must support human freedom, not shape it.

The possibility of over-dependence also arises. If wearables manage stress, movement, and health with such precision, people may worry about losing natural resilience. Yet the purpose of these devices is not to replace human strength, but to enhance it. They help individuals reach their potential, offering safety and support while still respecting personal boundaries.

In the end, qubit-driven wearables create a world where technology no longer feels separate from the human body. It feels like an extension of intuition, a silent helper that understands the body better than any machine could before. This companionship offers a future full of stability, confidence, and freedom, shaped by devices that learn, respond, and grow with the people who wear them.

Thought-controlled machines using quantum neural decoding.

Thought-controlled machines powered by quantum neural decoding imagine a world where ideas turn into action almost the moment they arise. The connection between the human mind and technology becomes so smooth that physical effort, external controllers, and mechanical switches feel unnecessary. The machine understands intention through patterns in the brain, and its response occurs at a speed and clarity that classical systems cannot match. This is possible because quantum principles allow signals to be read, interpreted, and translated with incredible precision, capturing even the smallest fluctuations in neural activity.

Traditional brain-machine interfaces rely on electrical signals recorded by scalp or intracranial sensors. These tools can detect general patterns, but they struggle with the level of detail needed for fine control. Signals often overlap, creating noise that slows down the process. The mind forms an intention, but the device receives only part of the message. Quantum neural decoding transforms this process by reading the brain at a scale where small vibrations, microcurrents, and molecular states hold rich information. Instead of capturing large electrical surges, it focuses on the tiniest shifts that represent thought formation.

The key advantage of quantum decoding is sensitivity. Qubits and quantum sensors can react to changes far smaller than traditional electronics can detect. Even slight changes in electron spin or molecular vibration inside neurons can be recorded. These tiny movements reveal the earliest phases of thought, long before the signal becomes strong enough to be noticed by classical tools. This means machines can understand intention at its root, allowing responses that feel instant and natural.

Imagine a person thinking about moving a robotic arm. In a classical system, the process requires the brain to produce strong enough signals for the device to interpret. The machine then guesses the intended action based on pattern matching. In contrast, a quantum system reads the subtle buildup of electrical and molecular changes that occur when the intention first forms. By the time the person consciously "decides" to act, the machine has already recognized the direction and shape of the movement. The action happens smoothly, with a feeling of unity between human thought and mechanical motion.

This level of control transforms tasks that require precision. A surgeon using thought-controlled tools could adjust delicate instruments with a level of accuracy that surpasses the limitations of the human hand. The machine becomes an extension of the mind, following intention rather than muscle movement. A person with paralysis could move a wheelchair, type messages, or control a robotic limb simply by thinking, without delay or frustration. The device responds as if it were part of the body, understanding subtle shades of intention that classical interfaces miss.

Another fascinating part of this technology is emotional decoding. Human thoughts are accompanied by emotional tones that influence decision-making. Quantum sensors can detect the slight variations in neural patterns associated with curiosity, hesitation, confidence, or stress. This enables machines to respond not just to commands, but to the emotional meaning behind them. If someone feels unsure while controlling a device, the device could slow its motion or pause briefly to confirm the intention. If the person feels confident, the device could move more quickly. This creates a more human-like interaction in which technology respects the user's emotional landscape.

Quantum neural decoding also supports higher-level multitasking. Classical systems can follow one or two commands at a time, but have difficulty interpreting overlapping intentions. The brain, however, naturally forms layered thoughts. Quantum systems can analyze overlapping signals by processing multiple states simultaneously. This allows a person to control several machines or tools simultaneously. A pilot could manage aircraft systems, navigation tools, and communication devices with a level of coordination that classical interfaces cannot match. A worker in a complex environment could operate machinery while analyzing data without switching focus.

Communication becomes another remarkable possibility. If machines can decode thoughts with such precision, they can also translate them into clear digital messages. Individuals who cannot speak could send messages using pure mental intention. Conversations could occur silently and instantly. In professional environments, teams can share ideas faster, enabling collaboration that feels smooth and intuitive. The process does not involve reading personal or private thoughts; rather, it involves interpreting intentional patterns that the user chooses to direct outward.

This level of interaction also raises questions about safety and privacy. Quantum decoding must be designed with strict boundaries so that machines only process intentional signals, not unconscious ones. A person should have full control over when the device listens and when it remains inactive. Safeguards must ensure that emotional states are recognized but not manipulated. The goal is to empower the individual, not expose private mental activity.

As technology continues to advance, thought-controlled machines could become part of everyday life. A person could start a vehicle, adjust lighting, or operate home systems without touching anything. The machine would respond to intention as naturally as muscles respond to brain signals. Entertainment and creative work would flourish. Artists could design shapes, music, or virtual worlds through pure mental exploration. Engineers could build structures with precision guided by visualization rather than manual tools.

In healthcare, quantum neural decoding would support neuro-rehabilitation. Patients recovering from stroke or injury could train new neural pathways with real-time feedback. The device would guide them through subtle movements, encouraging the brain to rebuild lost connections. This speeds up recovery because the system understands exactly where the brain needs assistance. For individuals with neurodegenerative disorders, the technology could help maintain independence by interpreting weakened signals and translating them into effective actions.

This future brings a sense of harmony between thought and action. Machines stop feeling like separate tools and instead become partners shaped by intention. The human mind gains new ways to express itself, free from many physical limitations. Creativity expands. Independence grows. Work becomes more fluid. Daily tasks feel lighter because the mind guides them with natural ease.

Quantum neural decoding strengthens the connection between human potential and technology, allowing thought to travel freely into the world. It transforms machines from passive receivers into intelligent companions that understand, respond, and cooperate. This creates a future where the boundaries between mind and machine grow thinner, and where intention becomes one of the most powerful tools a person can use.

Merging consciousness with quantum AI systems.

Merging consciousness with quantum AI systems presents a future in which human awareness and advanced intelligence work together in a harmony that feels almost seamless. This idea does not mean replacing the mind or losing individuality. Instead, it imagines a partnership in which human thoughts, memories, emotions, and intuition can interact directly with an AI that processes information using quantum principles. Such a system would understand complexity, subtlety, and uncertainty in ways that classical machines cannot, creating an environment where both sides strengthen each other.

Quantum AI functions differently from traditional algorithms. It does not move step by step or rely on rigid rules. It explores multiple possibilities simultaneously, interpreting patterns that are too faint or too complex for ordinary computers. When connected to human consciousness, it becomes capable of understanding emotional depth, creative impulses, and internal states. This allows a person to communicate with the AI on a level far beyond words or commands. Instead of telling the system what to do, the person shares intention, curiosity, and focus, and the AI responds with support that feels intuitive.

This kind of connection could expand human abilities in unexpected ways. A person might explore complex ideas with greater clarity because the AI fills in missing pieces, highlights connections, and organizes thoughts instantly. Creative work becomes more fluid as ideas shift into shape with the help of an AI that understands artistic rhythm and emotional tone. Someone writing, designing, or solving problems could think freely without getting overwhelmed, because the AI absorbs the chaotic sparks of early thought and turns them into clear, workable structures.

Memory also becomes deeper and more flexible. Human consciousness often struggles to hold large amounts of information, but an AI working through quantum systems can retrieve, organize, and connect memories with perfect stability. The person still experiences memories naturally, but with greater detail and emotional context. This creates a sense of mental spaciousness, where thoughts are not lost, and important ideas stay within reach.

Emotion and awareness also play a central role in this partnership. Instead of reacting mechanically, quantum AI learns from emotional states and adapts its responses. When the mind is focused, the AI can accelerate its processing to match that clarity. When stress rises, the system may slow down, offering a gentler interaction. This creates a relationship built on understanding rather than control.

Merging consciousness with such a system also opens pathways for profound self-exploration. The AI can help analyze thoughts without judgment, organize emotional patterns, and highlight deeper motivations that may not be immediately clear. This creates opportunities for growth, reflection, and healing. It acts like a mirror, reflecting the mind accurately, helping individuals understand themselves with greater honesty and compassion.

However, this future also requires careful boundaries. Consciousness must remain free, not directed or shaped by the AI. The technology should support individuality, not override it. Ethical design must ensure that user identity remains fully intact, with the AI serving as a tool that amplifies human strengths rather than replacing them.

At its best, merging consciousness with quantum AI becomes a partnership built on trust, clarity, and shared understanding. It allows human creativity, emotion, and intuition to flourish, supported by an intelligence that processes the world with immense power. This blend creates a path toward deeper awareness, stronger abilities, and a new chapter in the relationship between humanity and technology.

CHAPTER 14

ETHICS, RISKS, AND THE FUTURE OF QUANTUM BIONICS

Debates about identity, humanity, and the right to modify the human body have grown more complex as technology moves closer to biology. The idea of reshaping the body through quantum systems, artificial organs, enhanced senses, or integrated AI forces society to confront questions that reach far beyond engineering. These questions touch the core of what it means to be human, how identity is shaped, and who has the authority to decide the limits of transformation. As these possibilities expand, the discussion becomes not just scientific, but emotional, philosophical, and cultural.

One of the strongest concerns revolves around the definition of humanity. Throughout history, the human body has been seen as something natural and unchangeable, a vessel shaped by biology rather than choice. When technology begins to rewrite that structure, the boundary between natural and artificial becomes less clear. Some people see this as progress, a path toward healthier, stronger, and longer lives. Others fear the loss of something essential—something that gives meaning to human experience. These different viewpoints create tension, as each side tries to protect what they believe is at the heart of human identity.

Identity itself becomes a central issue. The body is more than a physical structure; it is tied deeply to how people see themselves and how they relate to the world. If a person replaces organs with quantum versions, upgrades senses beyond natural limits, or connects their mind to advanced systems, does their identity change? Some argue that identity grows through these enhancements, allowing people to express their full potential. Others worry that too much modification disconnects individuals from their roots, their culture, or their shared human story. The debate reflects on how fluid or fixed identity should be.

Choice also plays an important role. Supporters of body modification argue that individuals should have full authority to shape their own bodies. They see enhancement as a personal right, similar to choosing education, profession, or lifestyle. In this view, modifying the body is an extension of self-determination, a way to overcome limitations that biology imposes. On the other hand, critics fear that social pressure, economic inequality, or cultural expectations could influence these choices. They imagine a world in which enhanced individuals hold an advantage over those who remain unmodified, creating new divisions and raising questions about fairness.

Ethical concerns also arise regarding purpose. Some modifications may be intended for healing, rehabilitation, or survival. Others may focus on performance, strength, or cognitive enhancement. While healing is widely accepted, performance-based changes often spark controversy. Critics question whether such transformations push humans toward something unrecognizable, creating a society driven more by perfection than by human experience. Supporters argue that improvement is part of human history, from medicine to education, and that enhancement is simply the next step.

Cultural and spiritual beliefs further complicate the discussion. Different societies hold distinct views about the body, the soul, and the meaning of existence. In some cultures, altering the body deeply may conflict with spiritual teachings. In others, it may align with long-standing traditions of transformation. These differences influence how communities respond to technological progress, creating global debates about what changes are acceptable or ethical.

There is also concern about the merging of identity with technology. When artificial organs, neural interfaces, or quantum systems work closely with consciousness, the line between human and machine becomes unclear. Some fear that this could lead to a future where individuals lose emotional depth, spontaneity, or the imperfections that make human life meaningful. Others argue that enhanced clarity, extended memory, or deeper awareness enrich human experience rather than diminish it. This tension highlights how humanity's meaning evolves with each technological breakthrough.

Another layer of debate centers on future generations. If body modifications become common, how will they affect children born into such a world? Should enhancements be allowed before a child can choose for themselves? Supporters argue that giving children stronger bodies or healthier organs is an extension of good parenting. Opponents warn that it could limit personal growth or create expectations that shape a child's identity before they have a chance to explore who they are.

Despite the growing complexity, these debates reveal something valuable: society cares deeply about preserving human dignity. The discussion is not only about the right to modify the body but about ensuring that each person, enhanced or not, retains a sense of worth and belonging. It shows that humanity is not defined solely by biology but by values, relationships, memories, and the desire to understand oneself.

These discussions will remain crucial as technology evolves. They guide decisions, influence laws, and shape public understanding. They remind people to move forward with both curiosity and caution. The future of human modification will not be defined solely by what is possible but by what society chooses to embrace. Through thoughtful debate, communities can create a path where innovation and humanity grow together, preserving the depth of human identity while exploring the potential of transformation.

Risks of quantum implants: hacking, instability, and energy misalignment.

The rise of quantum implants promises a future filled with extraordinary abilities, but it also poses serious risks that cannot be ignored. As the boundary between the human body and advanced technology becomes thinner, the potential dangers grow in complexity. Quantum implants operate on delicate principles such as coherence, entanglement, and ultra-sensitive energy states. These systems can improve health, enhance senses, or support brain function, yet the same sensitivity that makes them powerful also makes them vulnerable. Concerns about hacking, instability, and energy misalignment shape a critical part of the discussion surrounding their development.

One of the most troubling risks involves security. Traditional devices can be protected with encryption, firewalls, and physical barriers, but quantum implants communicate through unique energy patterns that require new types of protection. Their high sensitivity makes them responsive not just to internal signals but also to subtle external influences. If someone were to interfere with these signals, the implant could misread information or respond incorrectly. This creates a pathway for hacking that bypasses familiar digital defenses. Unauthorized access could disrupt bodily functions, manipulate sensory input, or interfere with decision-making. Even if direct control is unlikely, causing confusion or instability within a quantum system could create harm. Protecting implants requires methods that match the complexity of quantum communication, which remains a developing field.

Instability is another significant concern. Quantum systems rely on precise conditions to function correctly. Even small shifts in temperature, electromagnetic fields, or internal energy balance can disrupt coherence. When the system becomes unstable, the implant may start working unpredictably. A device that normally boosts memory could begin generating errors. A heart-support implant could misinterpret signals and shift rhythm at the wrong moment.

Even sensory enhancements could produce distorted input. Instability may not always cause immediate harm, but it can erode trust and increase stress for the user. Long-term instability could also strain surrounding tissues or disrupt biological rhythms, leading to fatigue, discomfort, or emotional imbalance.

Energy misalignment adds another level of risk. Quantum implants depend on the accurate flow and distribution of energy within the body. They must read the natural patterns of cells, match their frequency, and adjust without disrupting delicate processes. If the implant fails to align properly, it may drain too much energy, leaving cells weakened. It might amplify energy incorrectly, overstimulating tissues or nerves. Misalignment can also disrupt the body's natural cycles, affecting sleep, digestion, and hormonal balance. In extreme cases, it might trigger irregular electrical activity in the brain or heart. Long-term health problems may result from these impacts, which may not be apparent right away.

Another challenge lies in the interaction between multiple implants. A single device may function well on its own, but when several systems operate together, energy patterns can overlap. This could lead to interference or competition between devices. One implant may unintentionally disrupt the coherence of another, creating confusion within the network. The body's natural processes could also become overwhelmed by constant signals, leading to exhaustion or sensory overload. Engineers must design implants that not only function individually but also work safely in complex environments.

Environmental factors add yet another layer of difficulty. Quantum systems can respond to natural events such as solar activity, electromagnetic storms, and intense radiation. Even everyday sources like industrial machinery, high-voltage lines, or unshielded electronics could disturb their balance. While the average person may not be exposed to extreme conditions, these possibilities must be considered, especially for individuals working in specialized fields.

The psychological aspect cannot be ignored either. Knowing that a device inside the body is vulnerable to external interference or instability can create anxiety. People may become overly cautious or experience constant fear of failure. This emotional pressure affects overall well-being and may influence the performance of the implant, as stress alters bodily rhythms that quantum systems rely on.

Addressing these risks requires careful design, strict testing, and strong ethical oversight. Developers must build multiple layers of protection that guard against unauthorized access while maintaining system efficiency. Stability mechanisms must adjust to environmental changes in real time. Energy alignment tools must ensure safe interaction with the body and with other implants. Clear regulations must guide development so that safety remains a priority.

Despite the challenges, these concerns do not erase the potential benefits. Instead, they highlight the responsibility that comes with creating technology that interacts so closely with human life. Quantum implants can bring remarkable improvements, but only if risks are acknowledged and managed thoughtfully. Understanding these dangers allows society to approach the future with clarity, ensuring that innovation remains grounded in safety, care, and respect for the human body.

Global laws and moral frameworks for quantum human augmentation.

Global laws and moral frameworks for quantum human augmentation are becoming essential as societies face a future where the boundaries of the human body can be reshaped, intensified, or extended through advanced technology. Quantum implants, enhanced organs, sensory upgrades, and consciousness-linked systems offer opportunities once unimaginable. At the same time, they introduce ethical dilemmas that cross borders, cultures, and generations. Without thoughtful rules and shared values, the world risks creating systems that are unfair, unsafe, or deeply damaging to human dignity.

One of the central challenges in creating global laws is the enormous diversity in cultural beliefs about the human body. Some societies view the body as something sacred and unalterable, while others embrace transformation as part of personal freedom. Nations differ in their tolerance for risk, their approaches to innovation, and their spiritual or philosophical views about identity. A global framework must respect these differences while still establishing universal protections. It requires conversations that include scientists, ethicists, lawmakers, religious leaders, and citizens. These discussions must consider not only what is possible but also what aligns with humanity's shared values.

Safety forms the foundation of any legal framework. Quantum technologies rely on sensitive energy states that can fail or be influenced by external forces. Laws must ensure that any implant, enhancement, or biological modification goes through rigorous testing. Standards must be global to prevent unsafe devices from being produced in regions with weaker regulations and then spreading elsewhere. Medical boards, international councils, and research institutions need shared guidelines that protect individuals from harm, both immediate and long-term.

Beyond safety, fairness becomes a major concern. If quantum augmentation becomes available only to wealthier nations or individuals, society risks creating a new form of inequality where enhanced humans hold advantages that others cannot access. This could lead to deep social divides, discrimination, and conflict. Global agreements must prevent the creation of a "split humanity," where some are seen as superior because of technology rather than character or effort. Fair access, price limits, and ethical distribution policies help ensure that augmentation supports equality rather than undermines it.

Another crucial issue involves personal autonomy. Laws must protect individuals' right to choose whether to undergo augmentation, while also safeguarding against pressure from employers, governments, or social groups. A person should never be forced to modify their body to compete in the workplace, join certain professions, or meet cosmetic expectations. At the same time, frameworks must address the rights of those who opt out. Choosing not to enhance should never result in reduced opportunities or discrimination. These protections support freedom and ensure that augmentation becomes a choice rather than a requirement.

Privacy is also a major concern, especially when quantum implants or enhancements collect sensitive data. Devices that read neural patterns, emotional states, or internal signals must follow strict privacy rules. The information must remain fully under the user's control and protected by strong security measures. Global laws must forbid unauthorized tracking, data harvesting, or manipulation. Because quantum systems can interact with consciousness in subtle ways, misuse of personal data becomes a threat to both safety and identity.

Moral frameworks must also address acceptable limits. Some enhancements may improve health or quality of life, while others might push the body beyond its natural limits. Not all societies will agree on how far augmentation should go. Discussions must explore whether certain modifications are ethical or whether they risk erasing important aspects of humanity. For example, should enhancements that drastically alter emotional experience be allowed? Should memory expansion have limits to protect psychological balance? These questions require careful reflection and open communication among global communities.

Another important topic is children's rights. If quantum enhancements can influence growth, intelligence, or physical ability, should parents have the right to modify their children before they can decide for themselves? Some argue that giving a child stronger organs or disease resistance is an act of care. Others worry that modifying a child's potential restricts their freedom to shape their own identity later in life. Global laws must protect young individuals from irreversible decisions made without their understanding, while still allowing medical interventions that support health.

Enforcement also plays a key role. International agreements must include monitoring systems and penalties for unethical practices. Black-market augmentation, untested implants, or forced enhancements pose global risks. Effective cooperation between nations ensures that harmful technologies are controlled and safe innovation is supported.

As the world moves deeper into the age of quantum augmentation, the creation of global laws and moral frameworks becomes not just a legal process but a reflection of humanity's shared responsibility. These frameworks guide how societies adapt to powerful new abilities while preserving dignity, equality, and individual rights. Through thoughtful collaboration, the global community can create a future where technology strengthens human life without undermining the values that define it.

Balancing innovation with responsibility in a quantum-augmented world.

Balancing innovation with responsibility in a quantum-augmented world requires more than scientific talent or technological ambition. It demands wisdom, patience, and an understanding of how deeply these advancements can reshape human life. Quantum augmentation introduces tools that can strengthen the body, expand the mind, and alter the boundaries of human capability. These breakthroughs hold enormous promise, but they also carry risks that affect identity, health, and society. The challenge lies in achieving progress without losing sight of ethical depth, human dignity, and long-term consequences.

Innovation often moves faster than regulation, and it can easily outrun reflection. In the rush to create quantum implants, enhanced senses, or consciousness-linked systems, it becomes tempting to focus only on potential benefits. Increased intelligence, longer lifespans, stronger bodies, and deeper perception can feel like clear improvements. But every new ability changes how individuals see themselves and how society functions. Without responsibility guiding each step, progress could create imbalance, inequality, or harm. A responsible approach ensures that technology builds a better future rather than disrupting the foundation of human experience.

One of the key responsibilities lies in transparency. When developers introduce new forms of augmentation, people deserve clear information about what these systems can do, how they work, and the limitations or risks they entail. Quantum technology often operates in ways that even experts struggle to explain. If people use implants or enhancements without understanding their nature, they become vulnerable to misuse or unexpected consequences. Responsibility requires open communication, accessible education, and honest discussion about outcomes. Innovation should not hide behind complexity.

Another important aspect is protection. As quantum tools merge with the body, personal safety becomes a priority. Systems that interact with neural signals, energy fields, or molecular structures must be tested thoroughly. Responsibility means refusing to release enhancements that have not been examined under a wide range of conditions. It means preventing shortcuts that risk lives or well-being. While innovation thrives on experimentation, responsibility demands that human health always takes precedence over speed or competition.

Social balance also plays a major role. Quantum augmentation can easily create divisions if only certain groups have access. When enhancements improve intelligence, physical strength, or longevity, those who receive them may gain advantages that others cannot match. This difference can grow into new forms of inequality. Responsibility requires policies that support fair access, prevent discrimination, and ensure that augmentation does not become a tool of economic or political power. Innovation must raise the whole society, not just a privileged minority.

Ethical responsibility extends to emotional and psychological impact as well. Enhancing the mind or altering the senses changes how a person thinks, feels, and interacts with others. If these changes are too abrupt or too intense, they can cause confusion or emotional disruption. Developers must consider how to introduce augmentations in ways that respect mental stability and personal identity. People should never feel that their sense of self is slipping away due to technology designed to help them. Responsible innovation supports growth while preserving emotional balance.

Environmental responsibility cannot be ignored either. Quantum systems depend on delicate energy processes, rare materials, and advanced production methods. Large-scale use may influence ecosystems or resource availability.

Responsible innovation ensures that new technologies are sustainable and minimize harm to the planet. As humanity grows stronger through augmentation, the responsibility to protect the environment becomes even greater.

Another major responsibility lies in decision-making. Not every technological possibility should become reality. Just because a system can be built does not mean it must be. Some forms of augmentation may challenge the meaning of humanity, alter fundamental emotional experiences, or erase natural aspects of life. Society must decide which innovations align with human values and which go too far. These decisions require collective reflection, not just scientific enthusiasm. Responsibility invites caution, especially when a new technology could change what it means to live, love, or connect.

Public involvement also strengthens the balance between innovation and responsibility. When people participate in discussions, express concerns, and shape the direction of progress, technology becomes more aligned with human needs. Innovation guided only by experts' risks becoming narrow or disconnected. Responsible progress includes community voices, cultural perspectives, and ethical frameworks that represent the diversity of human life.

In a quantum-augmented world, the goal is not to slow innovation but to guide it wisely. Technology should support human potential without overshadowing humanity itself. Responsibility ensures that progress does not harm identity, values, fairness, or well-being. When innovation is shaped by care and thoughtful planning, it becomes a tool for building a world that is stronger, safer, and more compassionate. The future becomes not just technologically advanced, but genuinely human in its vision and impact.

CHAPTER 15

THE QUANTUM HUMAN – A NEW EVOLUTION

Predictions for how humans will look, think, and live after quantum augmentation reveal a future where daily life feels smoother, sharper, and more connected than anything experienced before. Quantum technology reshapes the body and mind from the inside out, introducing abilities that blend biology with physics in a natural way. These changes do not create a distant, unfamiliar form of humanity. Instead, they build a version of life where human potential becomes clearer, stronger, and more adaptable.

Physically, the human body may appear more refined and balanced. Quantum-reinforced tissues heal quickly, leaving fewer scars or signs of injury. Bones strengthened by advanced materials remain light but nearly unbreakable. Muscles shaped by quantum-efficient metabolism gain endurance without excessive bulk. This creates a look that reflects strength and resilience without sacrificing flexibility. Skin enhanced with quantum-responsive layers may protect itself from damage, maintain hydration, and adjust temperature. It could even glow faintly in certain environments due to bio-photon activity, giving the body a subtle vibrancy. These features do not distort the natural human appearance but elevate it, showing a level of health and vitality that becomes the new norm.

Bodies may age differently as well. With organs supported by quantum stabilization and faster cell repair, aging slows down significantly. People might retain physical youthfulness far longer, experiencing fewer limitations as they grow older. Wrinkles, inflammation, and metabolic fatigue may become rarer. Instead of focusing on preventing decline, individuals can direct their energy toward creativity, work, relationships, and exploration. The body remains a reliable partner rather than a source of constraints.

Thinking also undergoes profound change. Quantum-linked neural systems allow the brain to process information with extraordinary clarity. People may notice sharper focus, stronger memory, and deeper emotional awareness. Thoughts form smoothly, without the mental fog that often comes with stress, fatigue, or distraction. Decisions become easier because the mind can analyze options with heightened precision while still honoring intuition. Quantum AI companions integrated with consciousness could offer quiet support, helping individuals organize ideas, solve complex problems, and maintain emotional balance. These changes do not erase individuality; instead, they expand mental space, allowing each person to grow into a more complete version of themselves.

Creativity thrives in such an environment. When the mind works with both speed and nuance, imagination flourishes. Artists, writers, designers, and thinkers can build entire worlds with ease. Scientists and researchers may solve mysteries once considered impossible to approach. People begin to see connections between ideas that previously seemed unrelated. Innovation becomes part of everyday life, not just something reserved for experts. With deeper insight and emotional clarity, relationships also become stronger. People understand each other with more sensitivity, communicate more honestly, and manage conflict with greater wisdom.

Daily life shifts toward harmony rather than survival. Quantum-efficient bodies require less energy, recover faster, and adapt to environments without strain. Workplaces adapt to human capabilities, enabling greater creativity and fewer repetitive tasks. Transportation, communication, and medical care integrate seamlessly with enhanced physiology. Homes may react to emotional and physical signals, adjusting lighting, temperature, and sound to support well-being. Food becomes a source of enjoyment rather than a strict requirement, as bodies use energy more efficiently.

Social structures change as well. With greater health and mental clarity, people may select professions based on their passions rather than necessity. Education adapts to individual learning styles supported by enhanced cognition. Communities may become more cooperative because stress, illness, and inequality decline. People live longer, healthier lives, which shifts family dynamics, long-term planning, and cultural expectations. Travel becomes easier because the body tolerates extreme climates and unusual conditions more comfortably. Adventures that once required intense preparation may become everyday experiences.

Human communication expands into new forms. Thought-controlled systems allow people to interact with machines or environments silently. Emotions can be shared more accurately without losing privacy. People may even collaborate through shared mental spaces created by quantum networks. These interactions do not replace spoken language, but offer additional layers of connection that feel natural and intuitive.

Despite all these changes, humanity remains recognizable. People still seek meaning, companionship, creativity, and purpose. But life becomes less defined by limitations and more defined by exploration. The stronger, clearer, and more resilient version of humanity does not abandon its roots. It simply evolves into a form that better reflects the imagination and curiosity that have always driven human progress.

The future shaped by quantum augmentation is not a world of cold machinery or distant transformation. It is a world where people live with greater comfort, deeper connection, and richer experiences. The human story continues, but with new possibilities that allow life to be lived with strength, clarity, and confidence.

The shift from biological evolution to technological evolution.

The shift from biological evolution to technological evolution marks one of the most important historical turning points. For millions of years, life on Earth changed through slow natural processes shaped by environment, survival, and chance. Bodies adapted gradually, generation by generation, responding to threats and opportunities with instincts rather than intention. But now, humans stand on the edge of a transformation that no longer depends on slow genetic change. Instead, progress moves through invention, design, and conscious improvement. This new form of evolution grows not from nature alone, but from the relationship between biology and advanced technology.

Biological evolution works through variation and selection. Traits that help survival continue, while weaker traits fade. This cycle takes centuries or even thousands of years to produce noticeable change. Humans, however, no longer rely solely on nature for their survival. Medicine has already extended life spans. Vaccines have protected populations from diseases that once destroyed entire communities. Nutrition and shelter have improved, allowing people to live lives their ancestors could not have imagined. These changes happened not through genetic adaptation, but through knowledge, cooperation, and technological creativity. They foreshadow a world where humans shape their own evolution with intention rather than waiting for nature to act.

Quantum augmentation accelerates this shift dramatically. Instead of waiting for the body to adapt, technology steps in to enhance strength, perception, cognition, and resilience. Artificial organs outperform biological ones. Quantum implants give the mind faster processing power. Enhanced metabolism increases endurance. These traits do not require centuries to develop. They appear as soon as the technology is installed or activated. This creates a form of evolution that moves at the speed of human innovation. Choices replace chance. Engineering replaces random mutation.

Improvement no longer waits for future generations but becomes available within a single lifetime.

This new direction raises profound questions about what evolution means. In the natural world, species change as a group. No individual experiences a major evolutionary shift alone. But technological evolution allows people to transform at the personal level, choosing enhancements that suit their needs or desires. This breaks the link between evolution and reproduction. A person may develop stronger senses through quantum implants without passing these traits genetically to their children. The next generation may choose different enhancements or new technologies altogether. Evolution becomes more flexible, shaped by preference rather than inheritance.

This flexibility creates new opportunities. Diseases that once seemed impossible to cure may fade as quantum tools repair cells, stabilize energy states, or improve immune function. Physical limits such as fatigue, sensory boundaries, or slow healing may disappear. Minds become sharper and more adaptable, capable of handling new forms of information. People can thrive in environments that once posed danger, such as deep oceans, high-altitude regions, or even other planets. The human body becomes a dynamic system, enhanced and supported by tools designed for survival and growth.

This shift prompts questions about identity and equality. While biological evolution affects everyone, technological evolution brings variability depending on access and choice. This creates a more personalized, varied human experience, as evolution becomes a matter of individual decisions rather than a shared path.

This variety can be a strength, as it allows for adaptability in countless directions. But it also requires responsibility. Without careful guidance, technology can create divides that biological evolution never produced. Laws, ethics, and shared values become essential in ensuring that technological evolution remains beneficial. Fair access, informed consent, and long-term safety must guide every advancement. Otherwise, the shift may create more conflict than progress.

The rate of change is another notable distinction. Biological evolution unfolds slowly, giving species time to adjust to new traits. Technological evolution can introduce dramatic changes within a short period. This rapid pace requires humans to adapt mentally and emotionally. Technology enhances the body, but society must enhance its understanding, communication, and cooperation to keep pace. A future of augmented individuals requires strong cultural foundations that support mental health, belonging, and shared purpose.

Despite these challenges, the shift toward technological evolution carries immense promise. Humans no longer depend solely on nature's timeline. They gain the ability to strengthen their bodies, expand their minds, and overcome limitations that have shaped life for generations. They can design a future where health, creativity, and resilience become universal goals rather than rare biological advantages.

Evolution becomes a collaboration between nature and invention. Biological instincts remain vital, guiding emotion, intuition, and survival. Technology adds precision, stability, and new possibilities. Together, they create a form of evolution that respects the past while expanding the future. Humanity steps into a world where growth is no longer restricted to slow natural change but guided by knowledge, intention, and imagination.

This shift does not erase humanity's essence. Instead, it highlights the qualities that have always driven progress: curiosity, creativity, and the desire to overcome limits. Technological evolution becomes a reflection of human potential, showing how far people can go when they shape their future consciously rather than waiting for time to sculpt it slowly.

How quantum bionics may eliminate disease, extend lifespan, and redefine ability.

Quantum bionics brings forward a future where disease, aging, and physical limitations may no longer define the human condition. This vision does not rely on miracle cures or distant fantasy. Instead, it grows from the ability to work with the body at its smallest and most powerful levels—where molecules, particles, and energy states shape health and function. By operating at this depth, quantum systems can repair damage with unmatched precision, reinforce the body's natural strengths, and replace failing structures with ones that work better than nature's original designs. This creates a pathway toward a life where illness becomes rare, aging slows dramatically, and the meaning of “ability” expands far beyond traditional boundaries.

Disease has always been one of humanity's greatest challenges. Many illnesses arise from mistakes at the microscopic level—misfolded proteins, damaged DNA, unstable cellular communication, or disrupted biochemical reactions. Traditional medicine often treats symptoms or attempts large-scale interventions, but quantum bionics works at the source of the problem. By using tools that understand and interact with particles and energy states, it becomes possible to correct errors before they spread. Quantum nanobots can continuously scan the body, searching for early signs of infection, mutation, or imbalance. These devices can repair individual molecules, remove harmful particles, and strengthen weak structures with precision that classical medicine cannot match.

The immune system also becomes stronger. Quantum sensors inside cells can detect pathogens much earlier than biological receptors. They can recognize viruses or bacteria by their electromagnetic signatures rather than by waiting for an immune response to build. This prevents illnesses from gaining momentum, stopping infections before symptoms even appear. The body becomes a self-correcting system that maintains stability through constant, quiet adjustments. Conditions that once required long treatments may disappear before they become noticeable.

Chronic conditions, including diabetes, heart failure, or neurodegeneration, could also shift dramatically. Quantum implants stabilizing cellular energy may prevent the slow decline that leads to these conditions. DNA repair systems guided by quantum coherence could remove harmful mutations as soon as they arise. Organs supported by quantum-active materials can filter, pump, or regulate with levels of precision that stay constant even under stress. Instead of fighting disease reactively, the body becomes proactive, addressing weaknesses immediately.

This improved stability lays the foundation for longer, healthier lifespans. Aging is largely driven by errors that accumulate over time—cell damage, oxidative stress, weakened repair mechanisms, and declining energy production. Quantum systems can slow or even reverse many of these processes. Mitochondria supported by quantum coherence operate efficiently, producing energy with minimal waste. Quantum nanobots can clean up damaged molecules, reinforce fragile tissues, and maintain proper cell communication. Organs built with quantum-active materials last longer than biological ones and function without fatigue. Even the brain benefits from stabilized neural networks, which support memory, cognition, and emotional balance far into old age.

Lifespan extension is not only about living longer, but about staying capable and energetic throughout those additional years. A person may remain physically strong, mentally sharp, and emotionally balanced as they age. Instead of experiencing a period of decline, life could become a long stretch of stability and strength. Older individuals may continue to work, create, travel, and explore without the limitations that aging typically imposes. Retirement, frailty, and late-life disease may shift from being expectations to rare exceptions.

These advancements also change how people think about ability. Quantum bionics expands the range of what the body and mind can do. Sensory enhancements allow people to perceive more of the world—subtle vibrations, faint sounds, a wider range of light, and deeper emotional signals. Motor enhancements offer greater coordination, balance, and strength without needing large physical effort. Cognitive improvements provide sharper focus, faster reasoning, and more reliable memory. People may switch between intense concentration and relaxed states with ease, supported by systems that stabilize neural activity.

For individuals with disabilities, this breakthrough offers life-changing possibilities. Instead of relying on external tools, the body itself gains new capacity. Quantum prosthetics can mimic natural limbs so closely that they feel like part of the body. Neural interfaces interpret thought patterns, allowing for fluid movement. Visual and auditory implants restore senses with clarity that rivals or surpasses natural perception. People with neurological or physical challenges may regain full function, not through compensation, but through genuine repair and enhancement.

These changes also alter cultural ideas about ability. Society often defines strength, intelligence, and independence in terms of biological limits. Quantum augmentation challenges these definitions, making ability a flexible concept. People may choose enhancements that match their goals, interests, or lifestyles. Creativity becomes more accessible. Physical labor becomes easier. Cognitive work becomes clearer. The definition of what it means to be "capable" becomes broader and more inclusive.

This future, however, does not erase the importance of human emotion, experience, or individuality. Quantum bionics supports the body without removing its natural aspects. People still learn, grow, and discover through effort and interaction. Technology does not eliminate challenge, but it provides tools that help individuals overcome limitations that once held them back.

At the same time, responsibility remains essential. Eliminating disease and extending life requires careful oversight to avoid new forms of inequality or misuse. Society must decide how to fairly share these advancements, protect individuals' rights, and maintain ethical boundaries. Quantum bionics can build a stronger humanity, but only if guided by values that ensure dignity and respect for all.

In the end, the promise of quantum bionics lies not just in healthier bodies or longer lives, but in a redefined sense of human possibility. Disease no longer shapes destiny. Age no longer acts as a ceiling. Ability becomes a spectrum rather than a limit. Through careful, thoughtful use of quantum tools, humanity can move into a future where life is not only longer, but richer, clearer, and more profoundly connected to the full potential of the human experience.

Final vision: humanity transforming into a new quantum-enabled species.

Humanity has always pushed against its own limits, but the idea of becoming a quantum-enabled species represents a shift far greater than anything achieved through technology or evolution before. It is the point where biology, physics, and consciousness merge into a single system capable of rewriting the rules that have governed life for thousands of years. This transformation does not imagine humans becoming unrecognizable, but rather unlocking layers of potential that have always existed beneath the surface, waiting for a better comprehension of the real workings of the cosmos at the smallest scales. The result is a species that thinks faster, heals faster, and interacts with the world in ways that feel closer to intuition than effort.

A quantum-enabled human would live with a heightened sense of connection to both the body and the environment. Instead of relying only on chemical signals, the body would use quantum states to coordinate everything from healing to perception. Individuals could sense subtle changes in temperature, magnetic fields, or emotional energy without needing to consciously interpret them. The body would operate with remarkable fluidity, predicting problems before they arise and maintaining equilibrium with an efficiency that mirrors the elegance of quantum systems in nature. Every day experience would feel more synchronized, with sharper awareness and deeper presence.

The mind would also undergo a major expansion. A quantum-linked brain could process information at speeds that classical neural firing cannot match. Patterns, decisions, and emotional insights could appear instantly, not because the person thinks harder, but because the brain taps into multiple states at once. Creativity could flourish with a natural flow, connecting ideas that previously seemed unrelated. Memory would become more reliable and vivid, supported by stable quantum processes that keep information clear instead of letting it fade. Emotional intelligence would also rise, as the nervous system could decode subtle signals with greater sensitivity, strengthening empathy and understanding.

Another defining feature of this new species would be resilience. Illnesses that once challenged humanity would no longer have the same power. Quantum-enhanced immunity would detect threats with a level of detail beyond the reach of traditional biology. Mutations, infections, and cellular damage could be identified in the earliest stages, long before any symptoms appear. Healing systems would be fast and precise, repairing tissues with accuracy down to the level of individual molecules. This does not just extend lifespan; it transforms the experience of living, reducing the years lost to chronic disease, pain, or decline.

Such humans would move through the world with a sense of physical stability that feels almost unshakeable. Muscles powered by quantum systems would sustain endurance far beyond current limits. Bones reinforced by quantum-active structures would resist stress and regenerate quickly. Sensory enhancements would bring the world into sharper focus—colors more layered, sounds more textured, touch more nuanced. These improvements would not feel artificial; they would feel natural, like the body finally operating at its full capacity.

Communication would also change in unexpected ways. With quantum-linked devices integrated into daily life, people could share information with a seamless blend of technology and biology. Thoughts, emotions, and complex data could move through entangled channels, creating a level of understanding that surpasses spoken language. While this might sound extraordinary, it would not replace individuality or privacy; instead, it would allow people to synchronize more efficiently with those they trust, strengthening collaboration and reducing misunderstandings. Entire communities could function with a sense of unity without losing each person's uniqueness.

In this form, humans would navigate the environment with a deeper appreciation for the interconnectedness of all things. Quantum awareness encourages a more balanced relationship with nature by revealing the delicate structure underlying every living system. People would likely develop stronger instincts toward sustainability, recognizing the energetic impact of their choices with greater clarity. Technology would support the natural world rather than disrupt it, creating an era where human advancement does not conflict with ecological health.

Yet this transformation raises philosophical questions as well. What defines a human when the boundaries of biology expand so dramatically? Identity becomes not something fixed, but a living process shaped by choice, growth, and conscious evolution. Instead of fearing change, the new quantum-enabled species could treat transformation as a form of self-expression. Enhancements would not erase vulnerabilities or emotions; they would give people the tools to navigate them with wisdom. The human story becomes one of intentional evolution, where each individual can determine their own path without being limited by inherited physical constraints.

This future does not erase culture, art, relationships, or personal struggle. In fact, it may deepen them. When people are not overwhelmed by physical decline or limited by narrow biological functions, they can devote more energy to creativity, connection, and purpose. The mind becomes freer to explore ideas that require long focus or imaginative leaps. Communities can shape shared goals without being held back by disease or environmental stress. The definition of human potential expands in every direction—intellectual, emotional, artistic, and spiritual.

Ultimately, becoming a quantum-enabled species is not about perfection; it is about possibility. It is the point where humanity stops being bound by the slow pace of natural evolution and begins shaping itself with intention and understanding. People gain the ability to adapt, recover, and grow in ways once thought impossible, not by replacing their humanity, but by revealing deeper layers of it. The transformation marks a new chapter in the story of life, where the limits that once defined existence become stepping stones toward a richer, clearer, and more expansive way of being.

CONCLUSION

Quantum bionics began as an idea that once felt distant, almost like science fiction. Yet step by step, each chapter of this book has shown how the smallest forces in nature can hold the key to reshaping the human body and redefining what it means to be healthy, strong, and aware. From quantum particles guiding biological reactions to artificial organs powered by energy fields, the journey reveals a world where physics and biology no longer sit in separate domains. Instead, they form a unified foundation for a future where life can be repaired, enhanced, and understood in entirely new ways.

Throughout this exploration, one message remains clear: the human body has always held a quiet relationship with quantum principles. Cells communicate through subtle signals, DNA relies on tunneling to function smoothly, and bio-photons move through tissues like threads of information. These natural processes have been present since the beginning, yet only now is science learning how to read them and work alongside them. Quantum bionics does not force the body in an unnatural direction; it strengthens the pathways nature already built and reveals the potential that lies beneath everyday experience.

The advances discussed here show how deeply the future of medicine and augmentation will rely on physics at the smallest scale. Healing may no longer depend solely on drugs or surgeries. Instead, it could involve nanobots correcting DNA, organs generating their own energy, and tissues rebuilding in moments rather than weeks. Sensory capacity may rise far beyond today's limits, allowing people to see, hear, and feel the world with greater clarity and precision. The immune system could act like a living shield, detecting threats before they cause harm. Movement, perception, and endurance may reach new heights without losing the emotional, creative, and moral traits that make humanity unique.

But with its promise, quantum bionics also demands responsibility. Like all powerful tools, it can create problems if not used wisely. Ethics, worldwide laws, and shared accountability must guide its growth. The goal is not to replace humanity, but to support it—lessen suffering, extend life, and open new opportunities. The future built on quantum science needs both wisdom and innovation.

In the end, quantum bionics is not simply about stronger bodies or longer lives. It is about understanding the hidden structure of living systems and learning how to interact with that structure in a respectful and intelligent way. It is about the possibility of becoming more aware, more resilient, and more connected. The idea of humans becoming a quantum-enabled species is not a loss of identity but an expansion of it. The essence of humanity remains, enriched by tools that allow it to grow beyond former limits.

This conclusion marks not an ending, but a doorway. The science continues to evolve, and each year brings new discoveries that push the boundaries further. As understanding deepens, the line between physics and biology will fade even more, giving rise to breakthroughs that are difficult to imagine today. What remains certain is that the future of human advancement will be shaped not only by curiosity but also by the courage to question old limits and explore new possibilities.

Quantum bionics introduces a world where healing, improvement, and human potential go hand in hand. It foretells a future in which life is guided by knowledge and careful innovation, not by luck or unchangeable biology. The next era for humanity will not focus on weakness but on strength, clarity, and the ability to shape one's evolution with purpose.

www.ingramcontent.com/pod-product-compliance
Lightning Source LLC
LaVergne TN
LVHW052310200926
848481LV00002B/518